S0-AGY-042

The World of Science

Facts On File®

AN INFOBASE HOLDINGS COMPANY

GLOSSARY

atom The smallest part of a chemical element that can exist alone and still be that element.

cell In biology, the basic building blocks of all living things. They are microscopic units of living matter, surrounded by a skin or membrane and containing jelly-like cytoplasm.

compound, chemical A substance composed of two or more elements joined chemically. For example, water is a compound of hydrogen and oxygen.

crystal A solid made up of blocks of a fixed, regular shape. Each grain of common salt is a crystal.

density The mass of a substance in relation to its volume, or mass per unit volume. It is usually stated as pounds per cubic foot, or kilograms per cubic meter.

electron A tiny electrically charged particle that circles the central nucleus of an atom. An electric current is a flow of electrons.

element, chemical A substance made up of one type of atom. Elements cannot be separated into simpler substances by ordinary chemical means. There are 92 elements in nature. All other substances are compounds.

force A push or pull that makes an object move, or change shape or direction. Examples of forces are gravity and magnetism.

frequency The number of times a vibration is repeated in a set time. The frequency of vibrations is measured in hertz (the number of vibrations per second).

friction The force that slows movement and produces heat when two surfaces are rubbed together. There is greater friction between rough surfaces than between smooth or oiled surfaces.

gravity The force that attracts any two bodies together. The larger an object, the stronger is its force of gravity. The large mass of the Earth creates a strong gravitational force, so objects fall downward toward the Earth.

hydraulics The transmission of forces by using liquids in pipes. A hydraulic ram uses liquids to transmit a large force to a piston.

lens A piece of glass or transparent material that has curved surfaces. Light passing through a lens is bent and can form an image. Lenses are used in cameras, microscopes and telescopes.

magnet An object that attracts iron and attracts or repels other magnets.
The magnetic force is strongest at two points called the north and south poles. A magnet (or an electric current) is surrounded by a magnetic field, or area where its magnetic effect can be felt.

mass The amount of matter in an object. Mass is different from weight because weight depends on gravity.

molecule The smallest particle of a chemical compound, consisting of one or more atoms combined.

neutron A particle found in the nucleus of all atoms (except hydrogen). Neutrons have no electric charge.

nucleus (1) In physics, the cluster of particles at the center of an atom. It contains protons and neutrons, about which electrons orbit. (2) In biology, a small dense body found in most plant and animal cells that controls the activities of the cell.

perpendicular at a right, or 90°, angle.

proton A particle found in the nucleus of all atoms. It has a positive electrical charge equal in size to the negative charge of an electron. It is about 2,000 times heavier than the electron.

wavelength The distance between successive crests of a wave. The wavelength of a light wave determines its color. The complete range of wavelengths for electromagnetic waves is called the electromagnetic spectrum.

The 101 Questions and Answers series contains six titles that cover a range of scientific topics popular with young readers, such as: the human body, geology, basic mechanics and physics, dinosaurs, and transportation. Each book is designed in a question-and-answer format with color illustrations throughout.

The World of Science

10 9 8 7 6 5 4 3 2 1

This book is printed on acid-free paper.

Printed in Italy

Library of Congress Cataloging-in-Publication Data
Lafferty, Peter.
The world of science.
p. cm. -- (101 questions and answers)
Includes index.
ISBN 0-8160-3219-X
1. Physics--Miscellanea--Juvenile literature. 2. Machinery--Miscellanea--Juvenile literature [1. Physics--Miscellanea. 2. Questions and answers.] I. Title. II. Series.
QC25.L24 1995
530--dc20 95-20019

Acknowledgments
Designer: Ben White
Editor: Lionel Bender
Media Conversion and Typesetting: Peter MacDonald
Project Editor: Veronica Pennycook.

Artwork credits
Art Beat-Ian Thompson: pages 4-5. Art Beat-Richard Dunn: 14-15 (top), 16-17, 20 (bottom), 28, 29 (top), 39, 42-43. Garden Studios-Darren Pattenden: 18-19, 20 (top), 22. Garden Studios-Roger Courthold: 6, 8-9, 23, 26-27, 40. Maltings Partnership: 10, 11 (top), 12-13, 24-25, 29, 30-38, 41, 42 (top), 43 (top). Mike Saunders: 7, 11 (bottom). Hayward Art Group: 14-15 (bottom), 37 (left), 44-47. Janos Marffy: 21.

CONTENTS

This book contains questions and answers on the following topics

Peter Lafferty

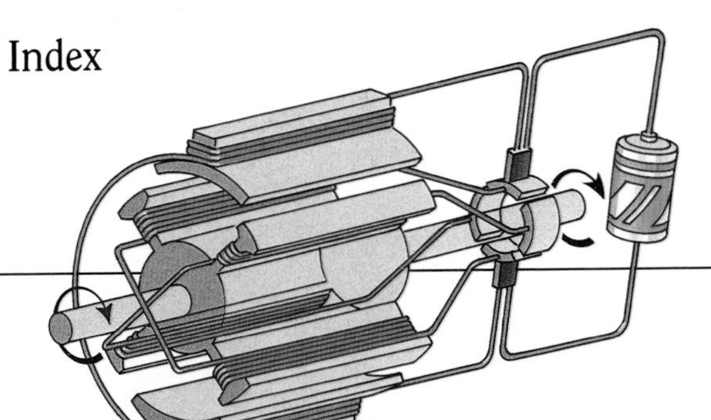

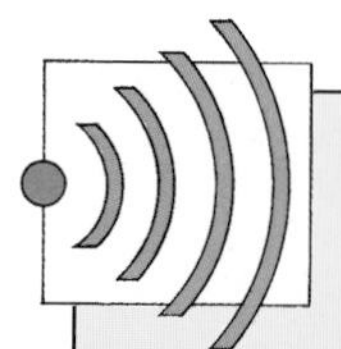

What is light?

Light is energy that travels through space as ripples of electric and magnetic force, spreading out like the ripples on a lake. The full range of these "electromagnetic waves" is called the electromagnetic spectrum. Our eyes are only sensitive to some of them, called visible light.

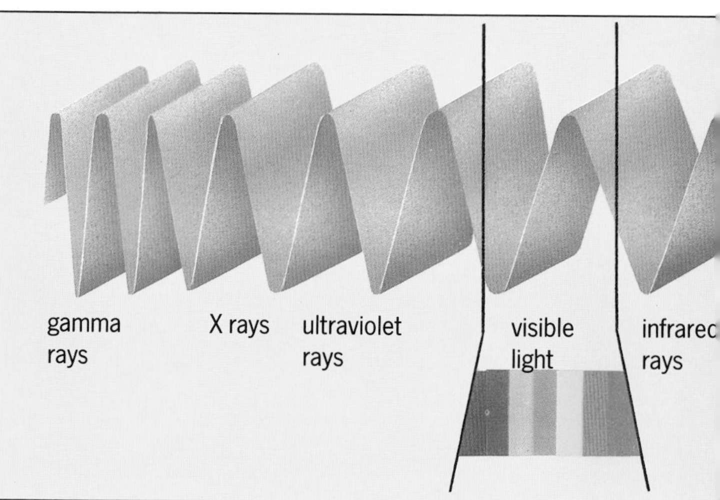

Why is a blackboard black?

Ordinary, or white, light is a mixture of the seven colors of the rainbow – red, orange, yellow, green, blue, indigo and violet. A blackboard appears black because it absorbs all the colors in the light falling on it. None are reflected (bounce off). But the white chalk used to write on the blackboard reflects all the colors, which our eyes combine to produce white.

How are colors mixed?

When beams of colored light are shone on a white surface, they mix to produce new colors. Any color can be produced by mixing different amounts of red, green and blue light. With paints, any color can be made by mixing combinations of red, yellow and blue paints.

Mixing colored lights

Mixing colored paints

HOW ARE COLORS SEPARATED?
A glass prism splits a beam of white light into the rainbow colors.

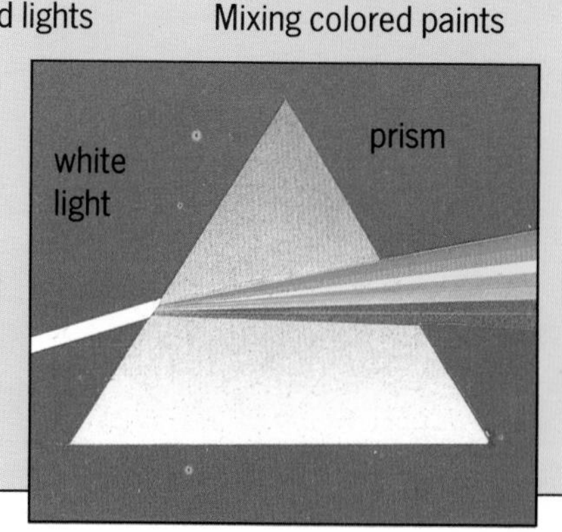

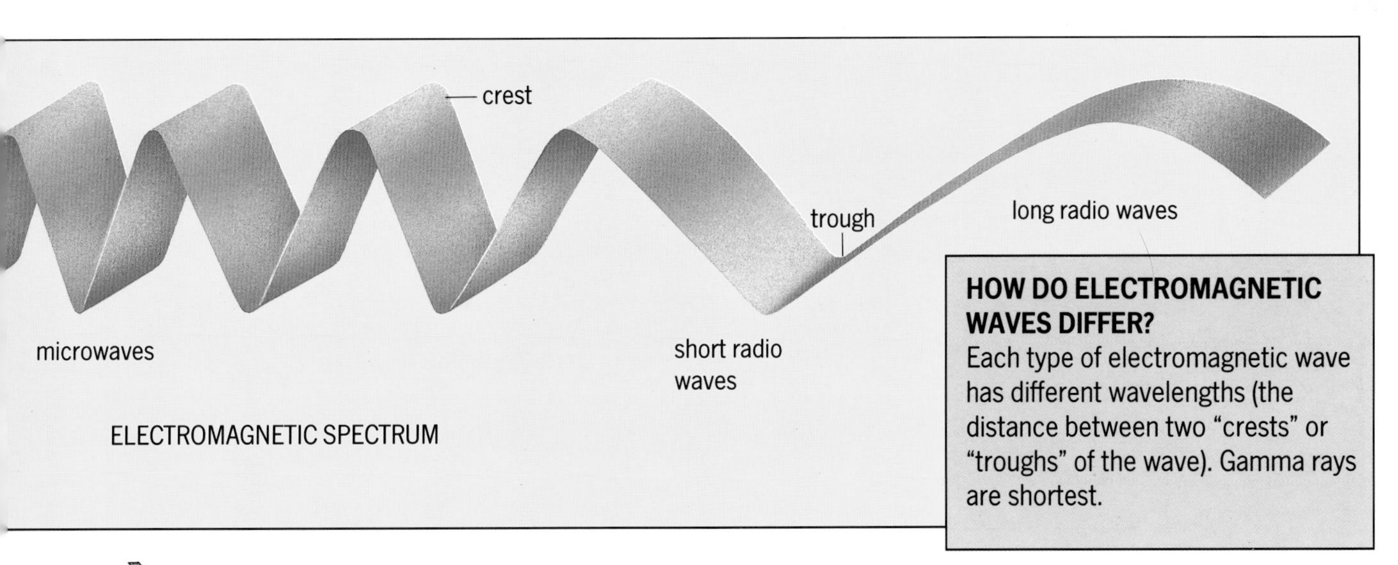

HOW DO ELECTROMAGNETIC WAVES DIFFER?
Each type of electromagnetic wave has different wavelengths (the distance between two "crests" or "troughs" of the wave). Gamma rays are shortest.

Why are things colored?

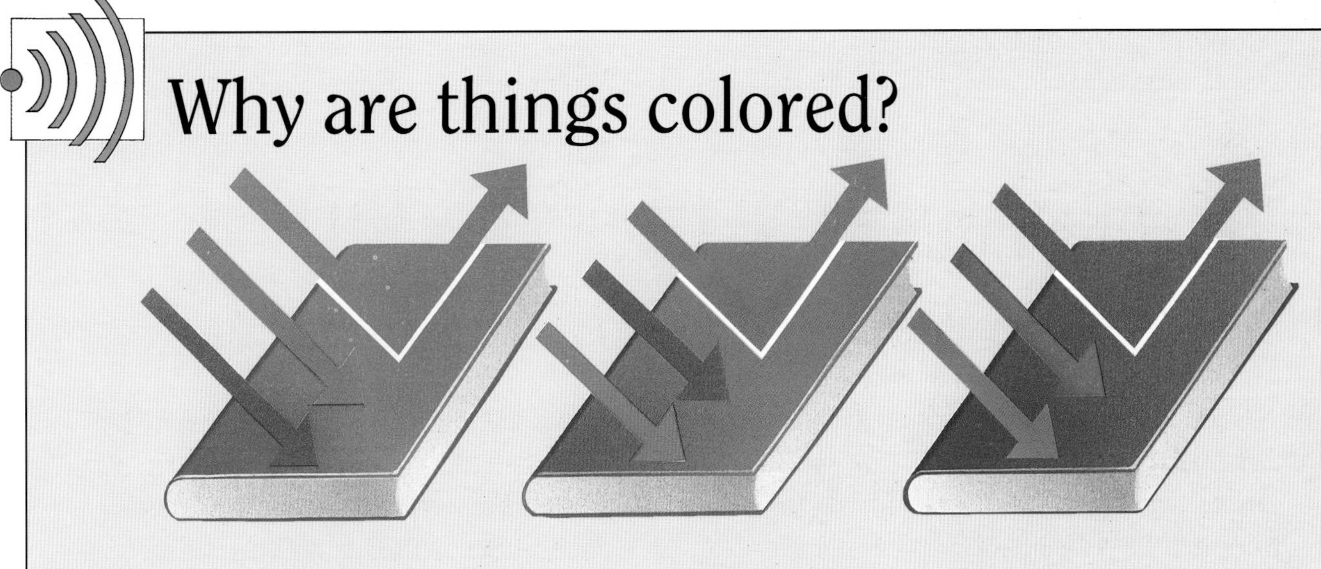

Different objects absorb different colors of the visible spectrum and reflect others. An object looks red because red is the only color that it reflects. It absorbs all other colors in the light falling on it. A green object reflects green light and absorbs the other colors. A blue object reflects blue light and absorbs red and green. A black object absorbs all the colors in the light falling on it. A white object reflects all colors. If a white object is viewed in red light, it reflects the red light and appears red. If a red object is viewed in green light, it appears to be black; there is no red light to reflect. A red object in red light is almost invisible.

HOW FAST IS LIGHT?
Light travels through empty space at a speed of 186,000 mi/sec (300,000 km/sec). Scientists believe that this is the fastest speed that anything can go. If you could travel at this speed, you would go around the world seven times in one second. The speed of light is a million times faster than a jumbo jet.

How does a mirror reflect light?

When a light ray is reflected by a plane (flat) mirror, it leaves the mirror at the same angle as it approached. When the light enters our eye we see an image of the object in the mirror. The image in a mirror is reversed; the right hand of a person looking into a mirror becomes the left hand in the image.

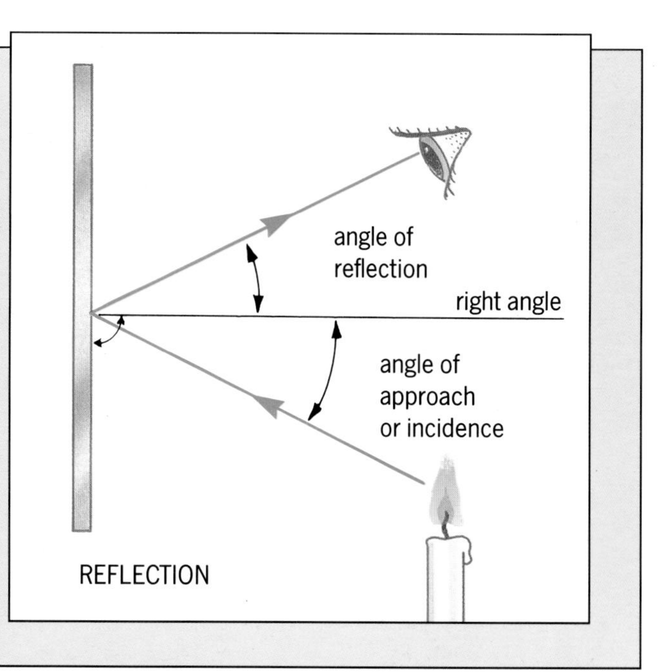

How does a periscope work?

A periscope lets you see around corners or look over walls. Inside the periscope, there are two mirrors tilted at an angle of exactly 45 degrees. Looking at the lower mirror, you see the scene reflected off the higher mirror.

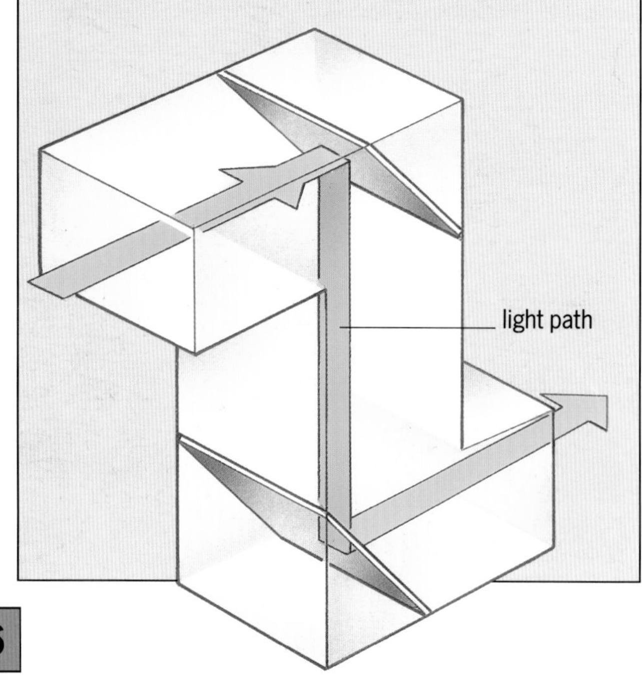

Are all mirrors flat?

No, some are curved. A concave mirror is shaped like the bowl of a dinner spoon. If you look into a shiny spoon, your image is upside down. The outer side of a spoon is another curved mirror; it is called a convex mirror. It produces an image the correct way up. Curved fairground mirrors produce strange images.

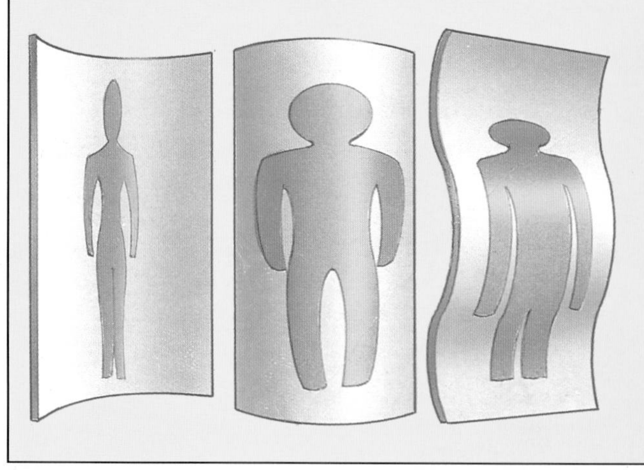

How is a mirror used?

Convex mirrors are used as rear-view mirrors in cars because they give a wide-angle view. Shaving mirrors are special concave mirrors. They produce a large, or magnified, upright image of a person's face.

Space colonies of the future, like the one shown here, could use large circular mirrors to reflect sunlight into the colony. The sunlight would allow plants to grow in the colony gardens.

ARE THERE MIRRORS IN SPACE?
In February 1993, the Russians placed a mirror in space to reflect sunlight onto parts of the Earth covered by night. In the experiment, an unmanned spacecraft unfurled a 63-ft (19-m) diameter disc of aluminum-coated plastic film. The experiment could lead to the placing of larger mirrors in orbit, reflecting sunlight over an area 10 mi (16 km) across. The mirrors would allow farmers to harvest crops at night.

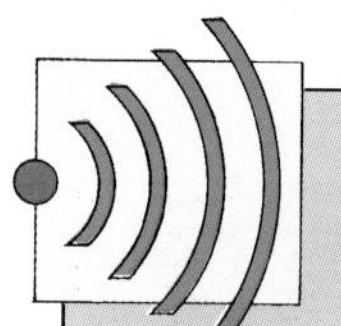

Can light bend?

Light travels slightly slower in transparent materials than through empty space. This causes a ray of light to bend slightly when it enters these materials. This bending of light is called refraction. When a ray of light travels from a less dense material to a more dense one, say from air into glass, the ray is bent toward the perpendicular.

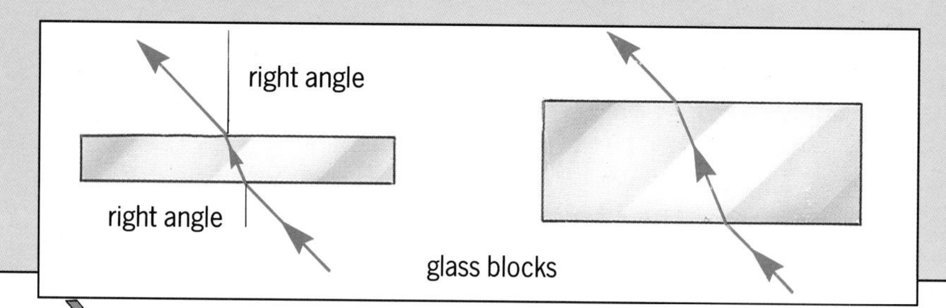

HOW ARE MIRAGES FORMED?

A mirage is the illusion of water seen on the horizon in hot countries. The mirage is due to the refraction, or bending, of light rays from the sky as they pass through the hot layers of air near the ground, so that they appear to come from the horizon. Since the light is from a blue sky, the horizon appears blue and watery. On cold nights, light can be bent in the opposite direction, so that objects below the horizon appear to float on it.

Why does a pool look shallow?

If we look from the side of a swimming pool, the water appears to be shallower than it really is. This effect is due to refraction. Light from the bottom of the pool is bent as it leaves the water and our eye is misled by the bent ray. We see an image of the bottom slightly above its real position. In the same way, refraction makes a stick floating half out of the water look bent at the middle.

HOW SLOWLY CAN LIGHT TRAVEL?

In ordinary glass, light slows down to 120,000 mi/sec (200,000 km/sec). Diamond slows it further to 74,400 mi/sec (124,000 km/sec). The lowest speed known for electromagnetic waves is 30,000 mi/sec (50,000 km/sec), in crystals of the element tellurium.

CAN A COIN APPEAR TO FLOAT?

When you look down at a coin in a glass of water, the coin appears to be much closer to the surface than it really is. This is due to refraction.

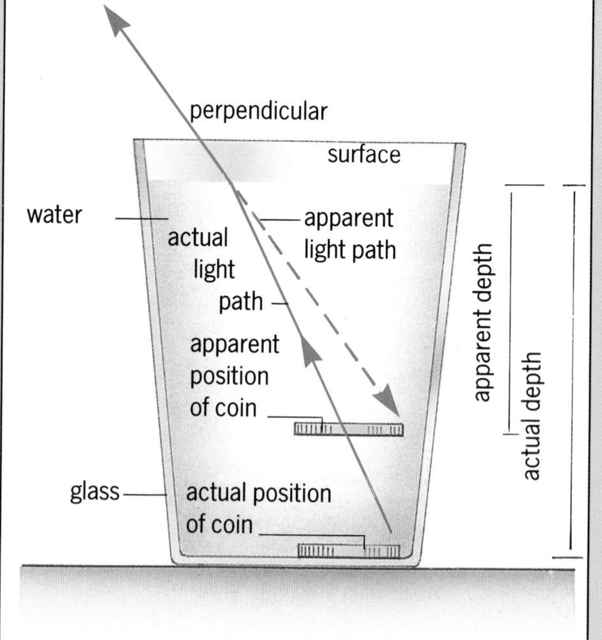

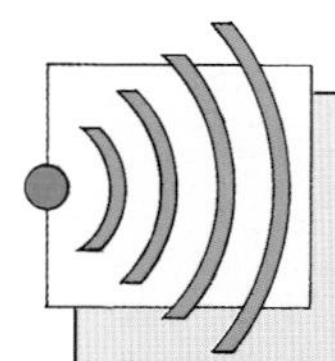

What is a lens?

A lens is a piece of transparent material with curved surfaces. Lenses bend light beams that pass through them. There are two main types of lens. Convex lenses bulge out at the middle. Concave lenses go in at the middle. Rays from a distant light source converge at a point called the focus after passing through a convex lens. A concave lens diverges light rays from a distant object, so that they seem to come from the focus.

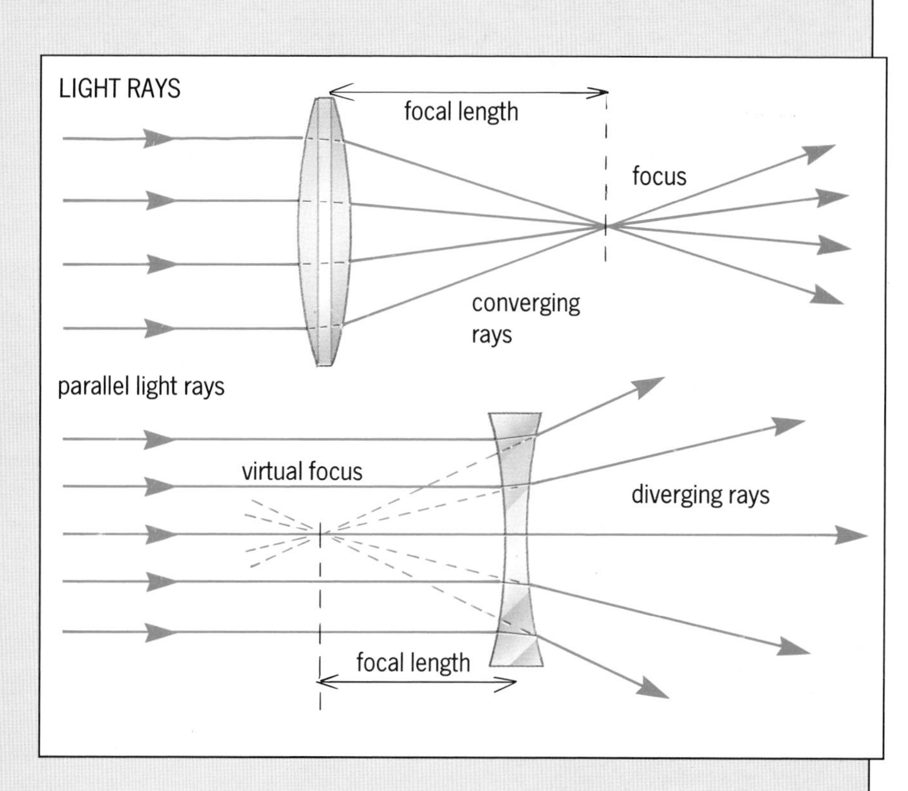

What are lenses used for?

Lenses are used in many optical instruments. Convex lenses are the most useful because they form images that are larger than the object being viewed. A magnifying glass is a simple convex lens. Cameras contain a convex lens that focuses light on the film. An image that can be focused on a film or screen is called a real image. A film projector produces a real image. The magnified image produced by a magnifying glass cannot be focused on to a screen; it is called a virtual image.

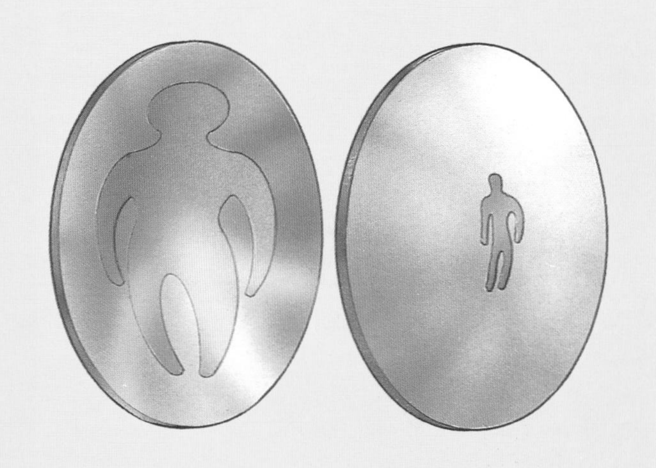

HOW BIG ARE LENSES?
Lenses can be almost any size. Small lenses used in microscopes are less than an inch across. The largest telescope lens ever made was 60 in (150 cm) across. It was shown in a Paris exhibition in 1900, but was flawed and was never used for serious scientific work.

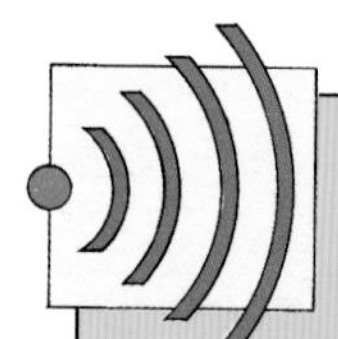

How does a microscope work?

A microscope uses lenses to produce highly magnified images. A compound microscope has at least two lenses. One lens (the objective) forms a magnified image, and then a second lens (the eyepiece) magnifies the image further, up to about 1,000 times. Each of them may be made up of several lenses to make the image clearer.

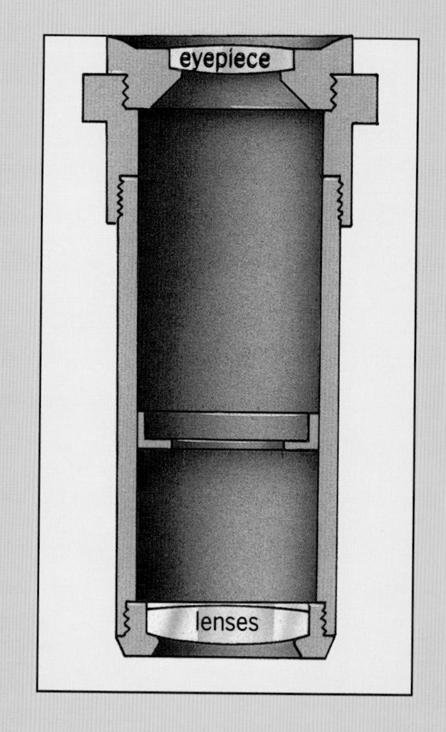

Here is a microscope with two sets of lenses. The specimen to be examined is placed on the slide, which is fixed to the stage. Light is reflected through the specimen using a mirror and lens (the condenser) beneath the stage.

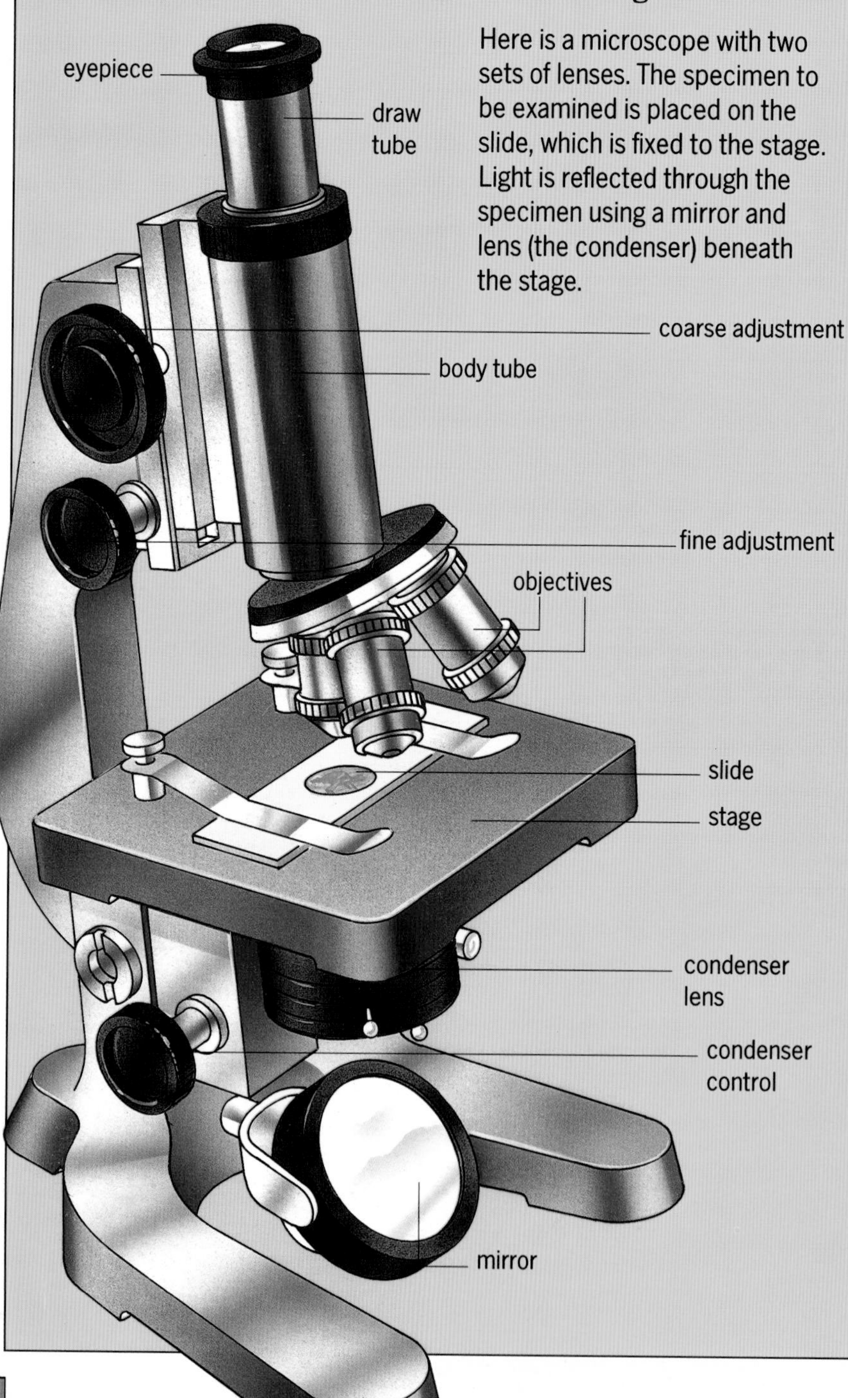

HOW IS A MICROSCOPE USED?
One of the objectives (below) is moved into place. Then the draw tube containing the eyepiece (above) is moved in and out to bring the specimen into focus.

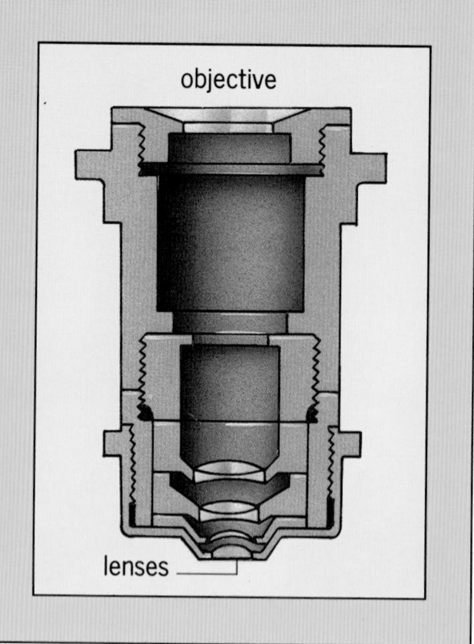

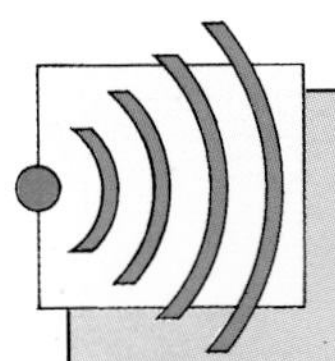

How do telescopes work?

Two different kinds of telescope are used by astronomers to view the night skies. The refractor uses two sets of lenses to gather and focus the light from the stars. The reflecting telescope uses a concave mirror to do the same job. The reflector illustrated (right) is called a Newtonian reflector. The concave mirror reflects light onto the plane mirror near the top of the telescope tube, and this mirror in turn reflects it into an eyepiece set in the side.

A large astronomical telescope (below) is housed in a dome, which has panels that open to let in the starlight.

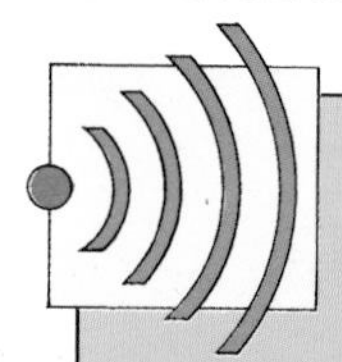

What is sound?

All sounds are caused by vibrating objects. As they vibrate, they cause ripples of high and low pressure in the air. The ripples spread out through the air as sound waves. Our ears pick up sound waves traveling through the air. With its huge ears, the desert fox has exceptional hearing.

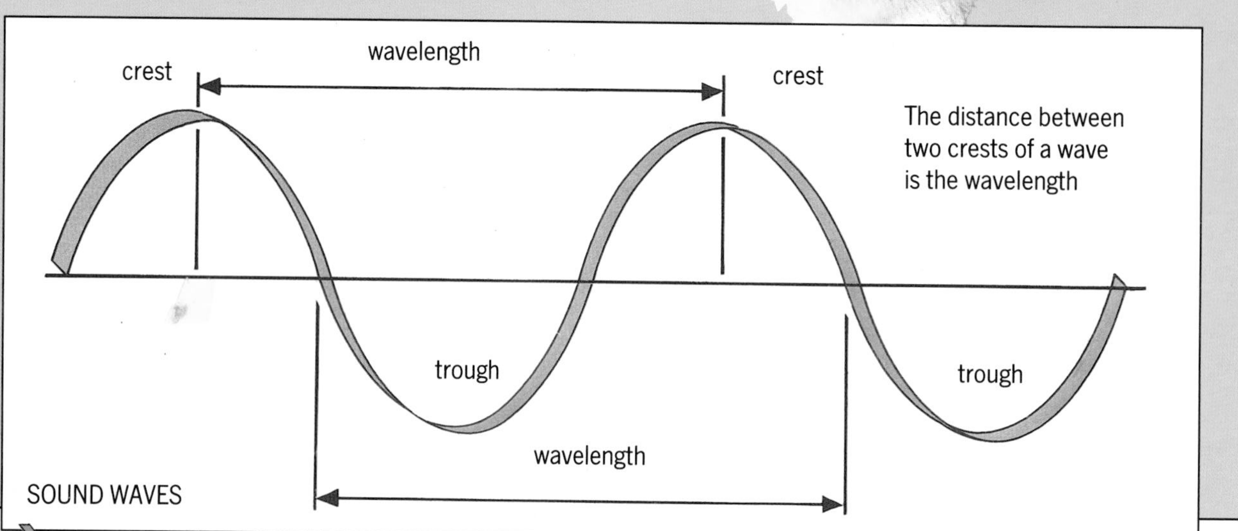

SOUND WAVES

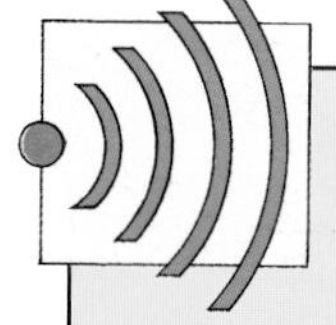

How can a rubber band make a sound?

Stretch a rubber band over a shoebox to make a simple guitar. Pluck the rubber band and it will make a sound as it vibrates. The vibrating rubber band pushes against the molecules of air in contact with it. The vibrations spread out through the air as waves, until eventually they reach our ears.

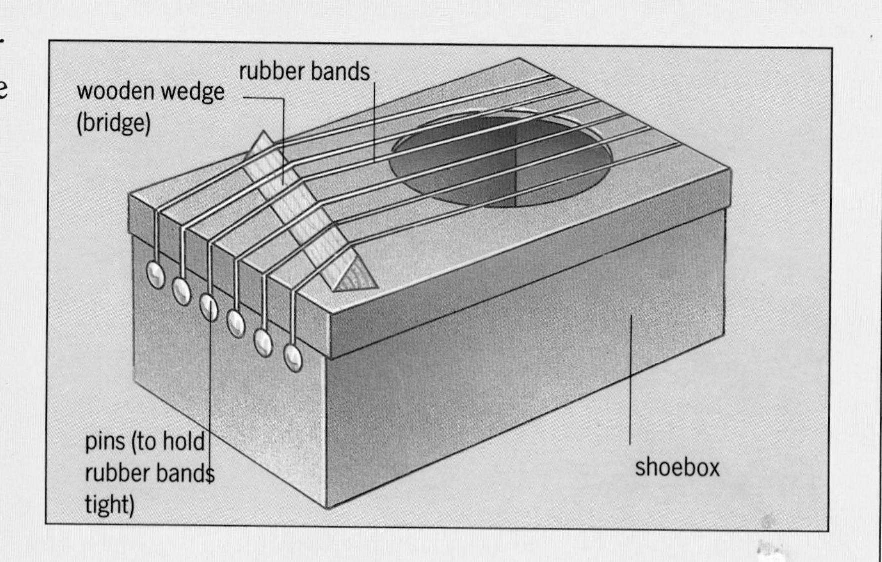

Why do sounds vary?

Sound waves can vary in amplitude (height of wave) and frequency (rate of vibration). Sound waves with a small amplitude are soft. In an orchestra, the quiet instruments are placed near the front, the loud ones to the back. Sounds with a low frequency are low-pitched.

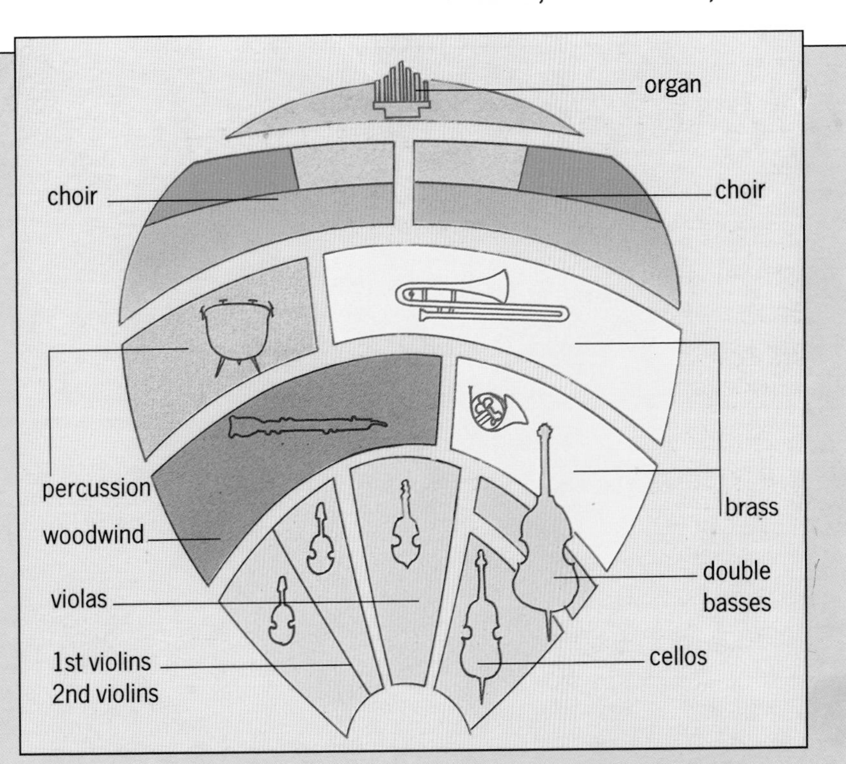

DID YOU KNOW...
Violins and cellos use vibrating strings and are called string instruments? Trumpets and organs use vibrating air columns and are called wind instruments? Drums vibrate when they are struck and are called percussion instruments?

WHAT MAKES MUSIC?
Some sounds that reach our ears are rhythmic and pleasant. Musical instruments are tuned to produce sound vibrations at the frequencies that we find most pleasant.

WHAT IS A NOISE?
Some sounds that reach our ears are irregular vibrations and are altogether unpleasant. We call them noise. The roar of a jet makes one of the loudest noises. A jackhammer is another offender. Continuous loud noise damages the ears.

WHAT ARE ACOUSTICS?
Sounds get softer the farther we are away from the source. Architects who design theaters and music halls try to make sounds carry to all parts of the theater. The Greeks and Romans built tiered amphitheaters in which to perform plays. These had remarkable sound qualities or acoustics. The hard seats did not absorb the sound waves, so the sound traveled without being weakened.

What is an echo?

An echo is a sound reflected off a hard object, such as a building. Ships use echoes to find the depth of the water they are sailing in. They are equipped with echosounders that bounce beams of high-pitched sound off the sea bed. The time the sounds take to travel between the transmitter and the receiver is measured.

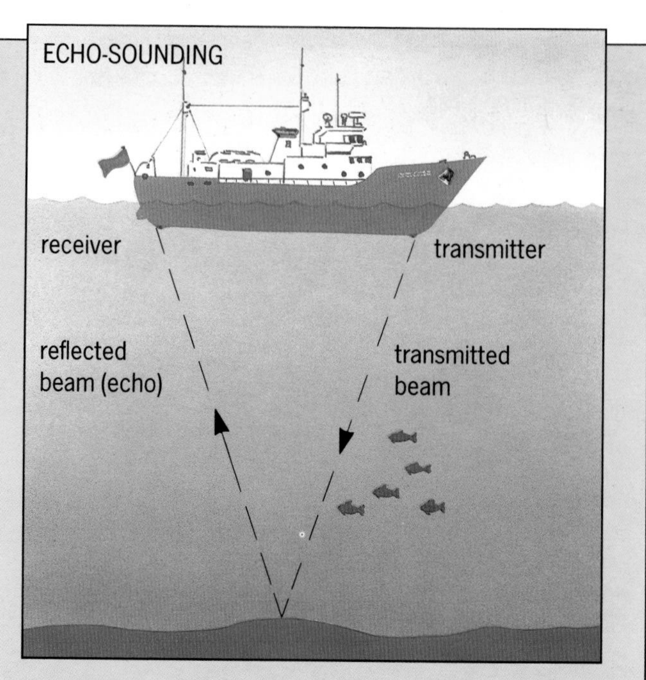

WHERE IS THE LONGEST ECHO?
The longest sounding echo in any building is said to be in the Chapel of the Mausoleum in Hamilton, Scotland. When the front door is closed, the echo of the slam lasts for 15 seconds.

WHAT IS THE SPEED OF SOUND?
Sound waves travel at about 1,089 ft/sec (330 m/sec) in air and at 4,950 ft/sec (1,500 m/sec) in water.

What is a sonic boom?

When flying below the speed of sound, a jet plane creates pressure waves that spread out at the speed of sound. As the aircraft reaches the speed of sound (Mach 1), the waves pile up on its nose. At supersonic (faster-than-sound) speeds, the aircraft overtakes the pressure waves, which form a cone-shaped shock wave. On the ground this is heard as a sonic boom.

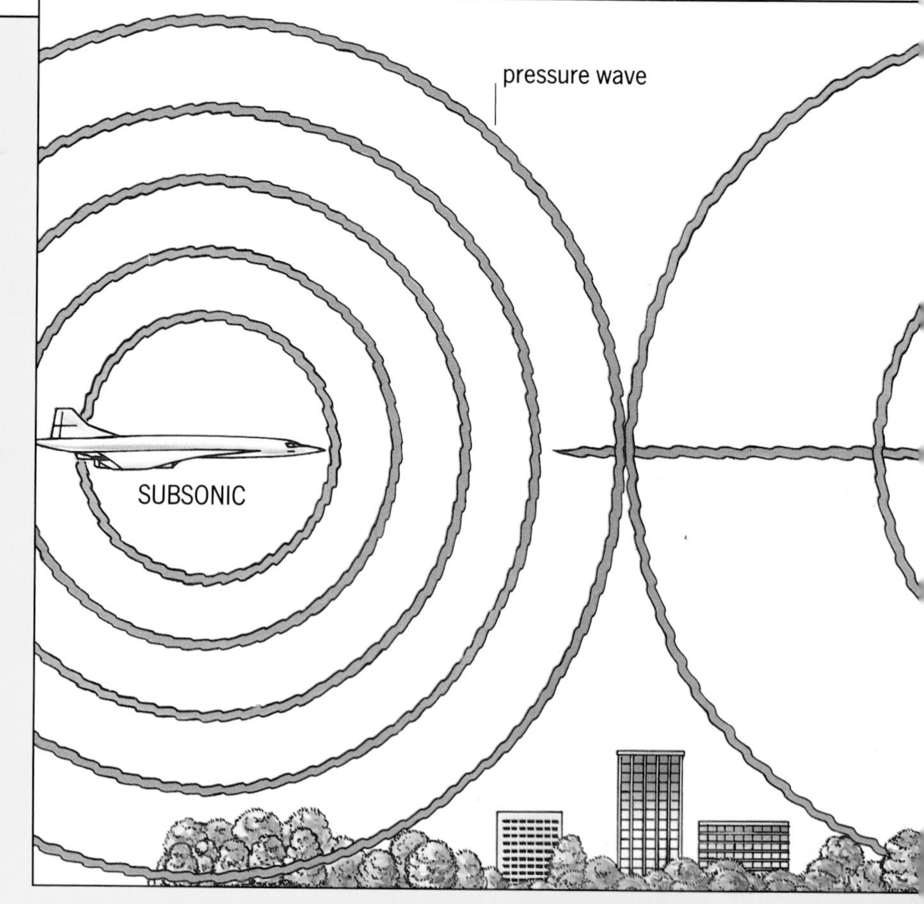

Why do bats squeak?

A bat locates its prey by emitting high-pitched squeaks (which we cannot hear) and listening for their echoes. For this kind of echo-location, bats use sounds with a frequency of up to 120 cycles per second (hertz).

DID YOU KNOW...

The first supersonic flight was made by Charles Yeager of the U.S. Air Force in 1947. Piloting the Bell XS-1 rocket-plane, he reached 647 mph (1,078 km/h) over a California airbase.

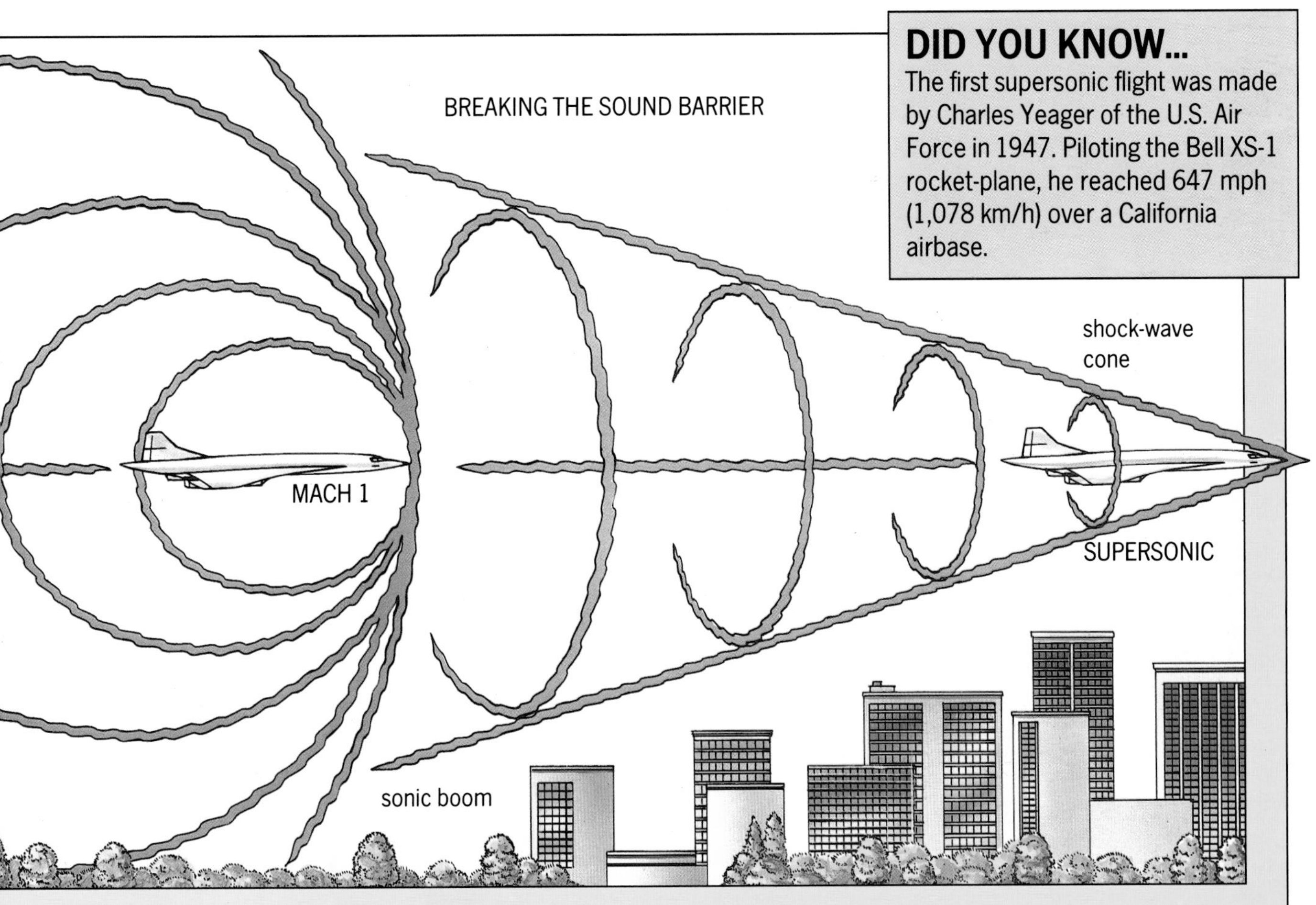

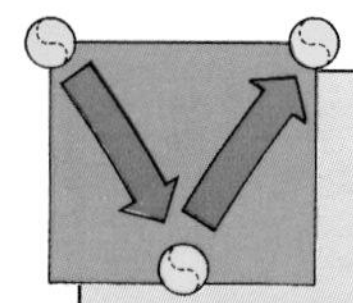

Why do things move?

When a soccer ball is resting on the ground, it will stay there unless you push it, pull it or kick it. If you do kick it, the ball will continue moving unless something stops it, such as a goalkeeper's hands. You could describe these two events by saying that an object (the ball) will remain at rest or continue moving unless it is acted on by a force (push, pull, kick). The English scientist Isaac Newton summed up the action of forces on matter in a similar statement more than 300 years ago. It is called his First Law of Motion.

HOW IS A FORCE MEASURED?

Forces are measured in units called newtons (N). Here are some examples of strengths of forces:

- force on a grasshopper's legs when it jumps – 0.001 N
- force when you cough against your hand – 0.01 N
- force needed to peel a banana – 1 N
- force of the racket hitting a tennis ball – 10 N
- force used to kick a ball – 100 N
- force of a car engine – 1,000 N
- force of a train engine – 10,000 N
- force of a rocket engine – 1,000,000 N.

WHY DO THINGS ACCELERATE?

When a motorist pushes the car's accelerator, the engine produces a greater force. This makes the car speed up, or accelerate. The greater the force on an object, the more it accelerates. This is called Newton's Second Law of Motion.

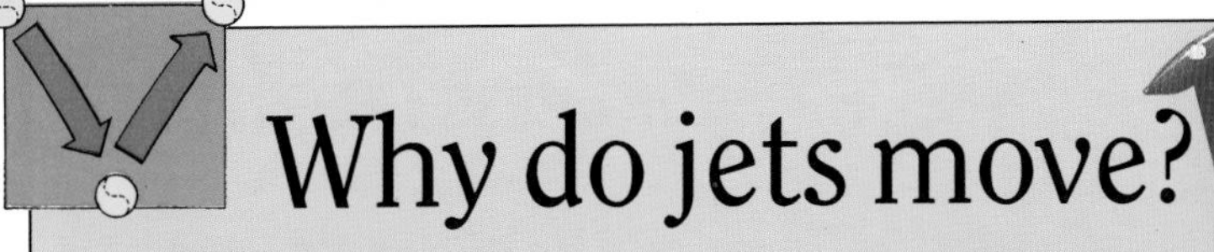

Why do jets move?

A force applied in one direction gives rise to an equal force in the opposite direction. This statement is called Newton's Third Law of Motion. If you are on roller skates and push against a wall, the wall pushes you away. When fuel burns in a jet engine, it produces hot gases that shoot backward. This gives rise to a force forward, which propels the jet plane.

What is friction?

A ball rolling across the ground will soon slow down and stop because a hidden force is acting upon it, the force of friction. This force is caused by the ground rubbing against the ball. Friction exists between any two surfaces rubbing together. When the Space Shuttle returns from space, friction between the air and the craft slows it down.

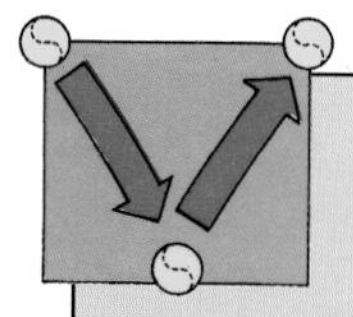

What is weight?

When an apple is ripe, it falls from a tree. It falls because it experiences a force called gravity. Gravity is a pulling, or attractive, force that any large object like the Earth exerts on everything both on and around it. The Earth's gravitational pull gives an object its weight. A pair of equal-arm scales is used to compare weights. The arm balances when the weights in both pans are identical, as shown here.

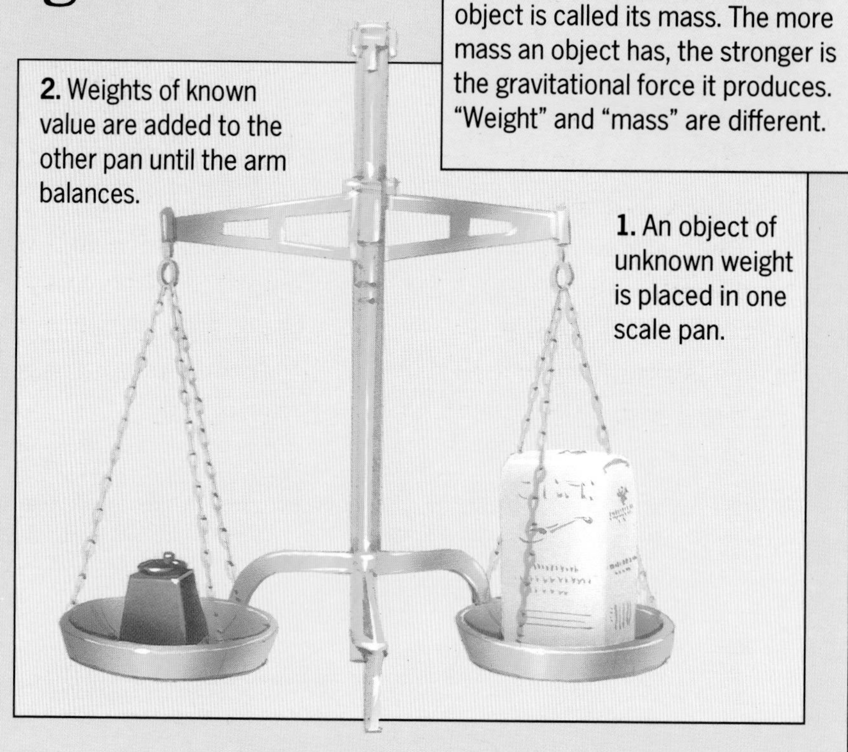

2. Weights of known value are added to the other pan until the arm balances.

1. An object of unknown weight is placed in one scale pan.

WHAT IS MASS?
The total amount of matter in an object is called its mass. The more mass an object has, the stronger is the gravitational force it produces. "Weight" and "mass" are different.

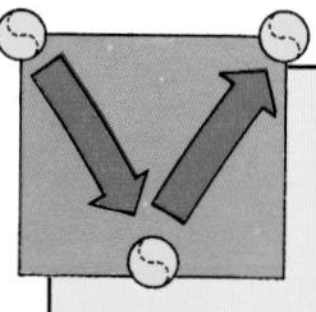

How far can you throw?

Gravity keeps our feet firmly on the ground and gives us our sense of up and down. Down is the direction in which things fall when they are dropped; up is the opposite direction. We can temporarily overcome gravity by movement, say, by throwing something into the air. But it doesn't go far before gravity pulls it back down to Earth. The faster you throw the object, the farther it goes. If you could throw an object at a speed of 16,800 mph (28,000 km/h), the object would circle round the Earth without falling.

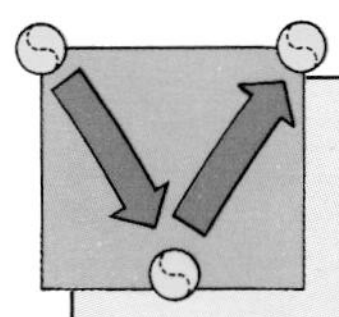

What is density?

A stick of butter and a feather-filled pillow may both have the same mass. They have a different size, however, because they have a different density, or mass per unit volume. The butter is smaller because it has greater density.

The Moon is smaller and less dense than the Earth. Its surface gravitational pull is six times weaker than Earth's. On the Moon, your weight would be one-sixth of your weight on Earth (although your mass would be the same).

HOW FAST CAN YOU FALL?

Falling objects speed up, or accelerate, as they fall. Without air resistance, a falling object would reach a speed of 60 mph (100 km/h) in less than 3 seconds. However, air resistance slows falling objects. Skydivers, for example, reach a maximum speed called the terminal velocity of about 180 mph (300 km/h). They fall at this speed until they open their parachutes.

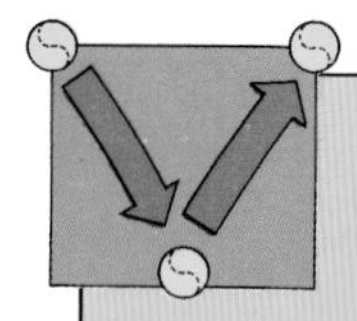

How do satellites stay in orbit?

When you whirl an object on a piece of string, you can feel the object is pulling outward and about to fly off. To keep it traveling in a circle, you have to pull inward on the string. Similarly, when a satellite is circling (orbiting) the Earth, it is trying to pull away. But the inward pull of gravity stops it from flying away.

Why don't you fall from a roller coaster?

When a car swings around a curve, the passengers feel a force that pushes them outward, away from the center. On a roller coaster, the car is continuously turning and the same force pushes the passengers against their seats so that they do not fall out.

WHAT IS CENTRIPETAL FORCE?
It is the force acting toward the center of a circle to keep an object moving in a circular path. It balances the outward-pushing force, called centrifugal force.

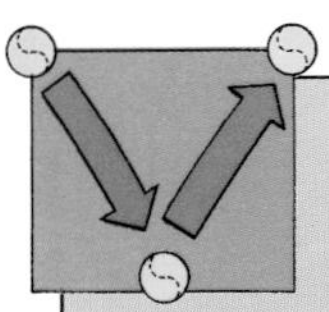

Do astronauts fall?

Yes. In spacecraft in orbit, astronauts are always falling! But, at the same time, they are moving forward at great speed. The combination of the two movements produces a circular path that keeps them at the same distance above the Earth, so they never crash to Earth. The astronauts appear to be weightless.

WHY ARE TELEVISION SATELLITES STATIONARY IN THE SKY?
Television satellites appear to hover in one fixed position in the sky. This is because they orbit at a height of 22,300 mi (35,900 km) over the equator. At this height, they take exactly 24 hours to circle the Earth. During this time, however, the Earth also turns around once and so the satellite appears fixed in the sky. Such an orbit is called a geostationary orbit.
As well as television satellites, communications satellites, which relay telephone, fax and computer messages, are placed in geostationary orbits.

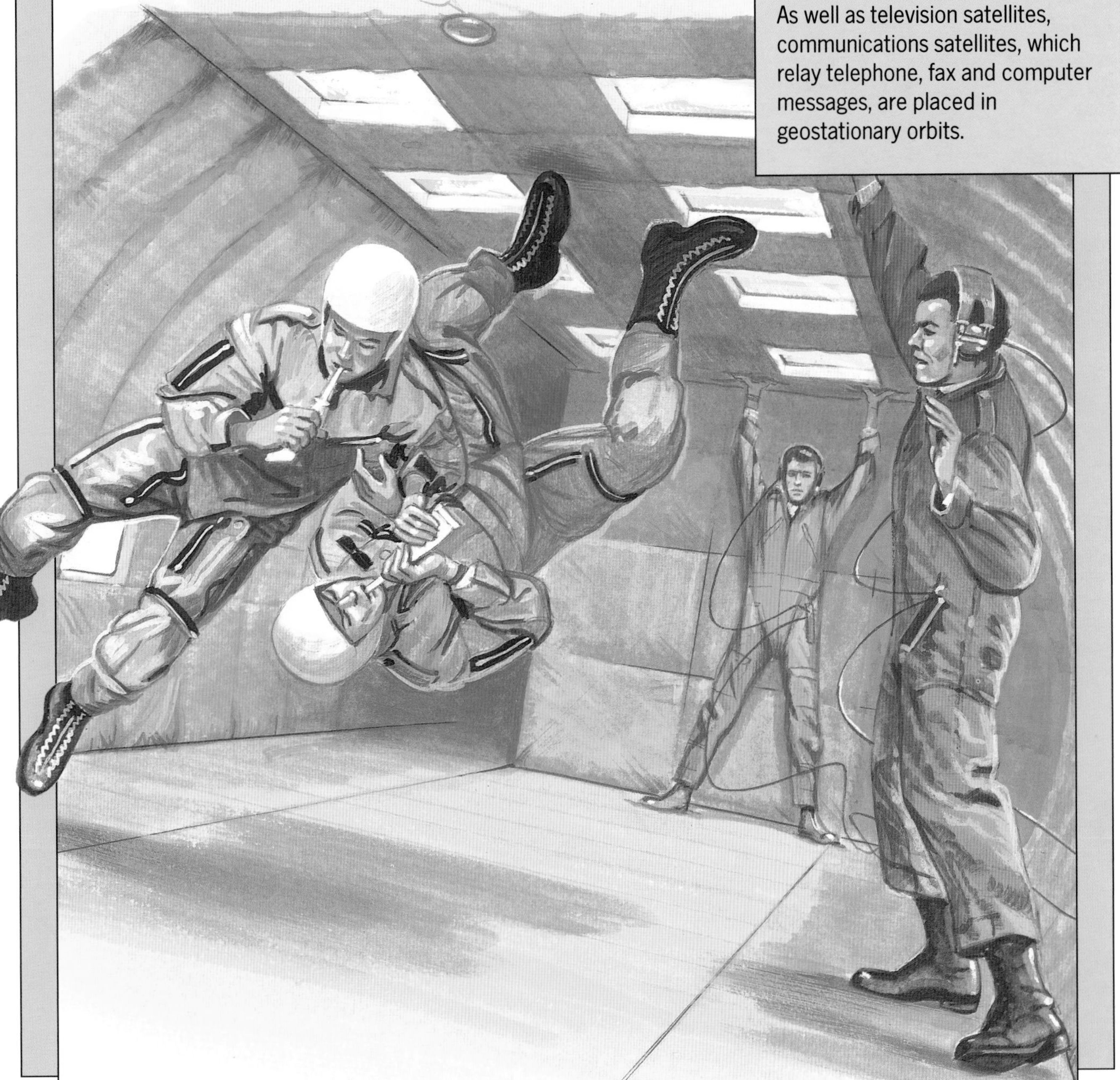

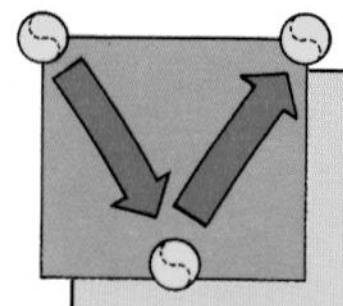

What is static electricity?

Sometimes when you comb your hair with a plastic comb, the comb and your hair become electrified. Your hair and comb have acquired different electric charges (+ and −), and attract one another. What happens when the electrified comb is held near a stream of water from a tap?

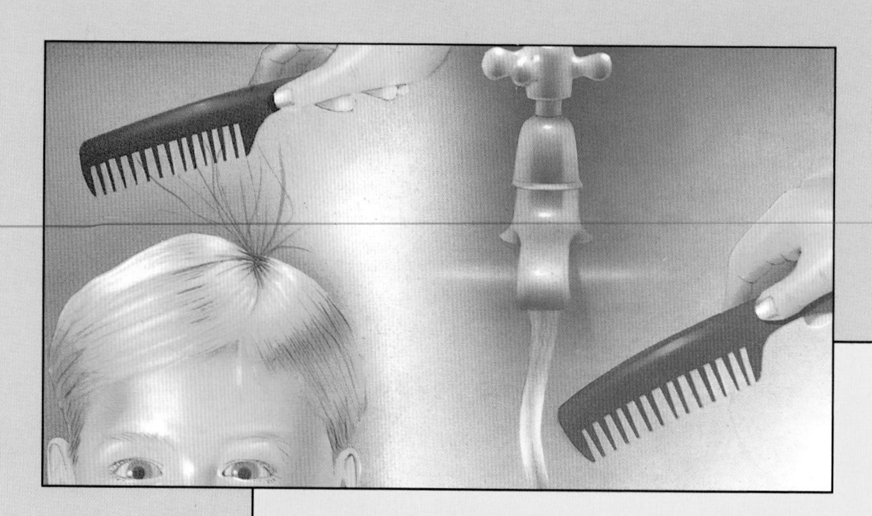

What is lightning?

The rubbing together of droplets in a rain cloud produces very large electric charges. When this electricity discharges, we see the great sparks we call lightning. You can produce electric sparks, on a smaller scale, by rubbing a balloon against your sweater. Two balloons with the same type of charge will repel each other. A charged balloon will stick to a wall.

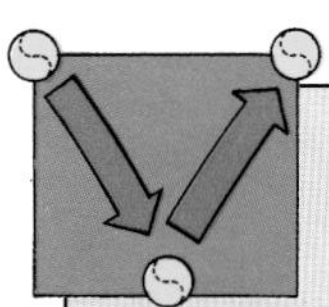

What is a magnet?

A magnet is a piece of iron or steel that can attract other pieces of iron and steel. Iron minerals such as lodestone are natural magnets. Sailors once used pieces of lodestone as compasses. When a magnet is suspended, it comes to rest pointing north-south. We call the ends of a magnet north and south poles according to which direction they point in.

DO MAGNETS ALWAYS ATTRACT?

No; it depends on which poles are close together. Opposite magnetic poles attract one another. Similar poles repel, or push apart.

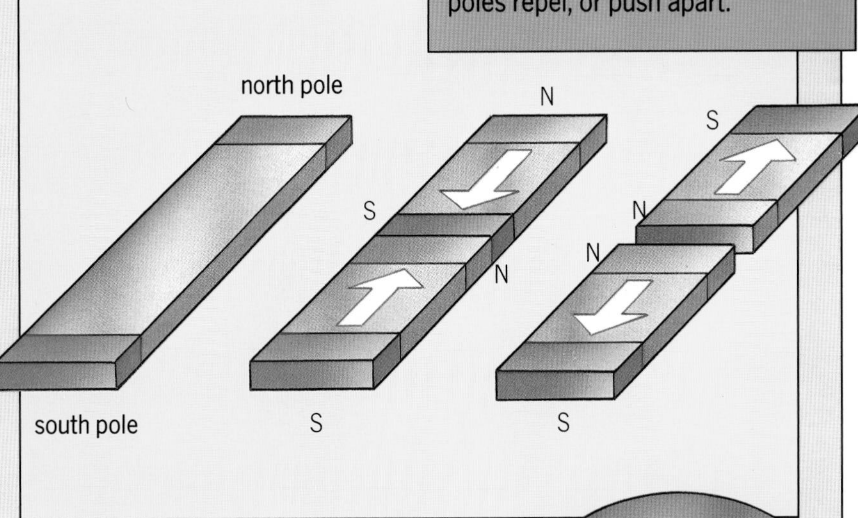

Horseshoe magnet

S

pins

N

WHAT ARE LINES OF FORCE?

Iron filings arrange themselves into patterns called lines of force, when they are sprinkled around a magnet. The curves follow the direction of the magnetic force, starting from, or leading to, the poles of the magnet, as shown below.

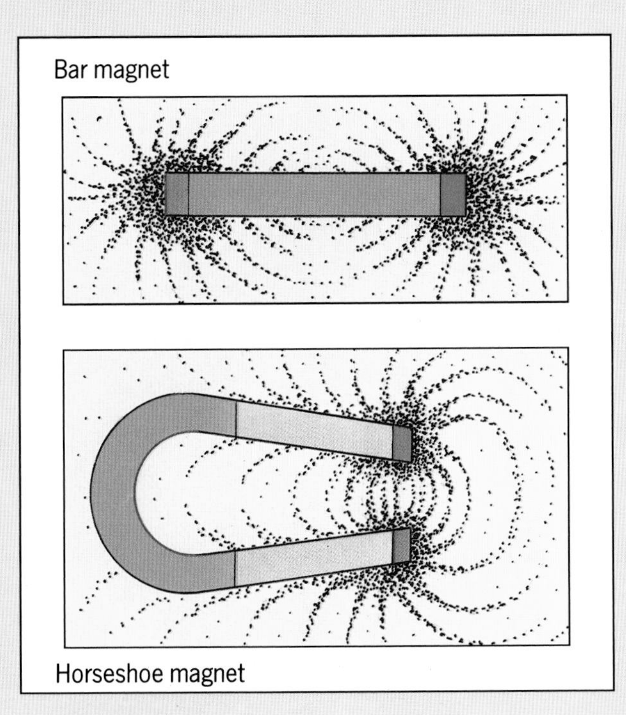

IS THE EARTH A MAGNET?

A horseshoe magnet attracts steel pins. The region around a magnet where magnetic forces act is called the magnetic field. The Earth is like a huge magnet. It has a magnetic field that makes magnets and compasses point north-south. The Earth's magnetism is thought to be caused by currents of molten iron moving in its central core.

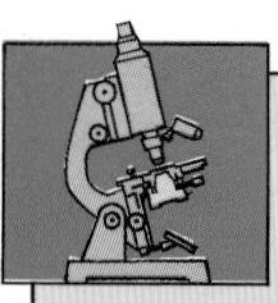

What is an atom?

All materials are made up of tiny particles called atoms. Atoms are very, very small – about a 400th of a millionth of an inch across. About 100 billion atoms would fit on the period at the end of this sentence. Each chemical element has one type of atom. The properties of the element depend on the way its atoms behave and, in particular, how they form molecules (see pages 25, 30). The word *atom* comes from the Greek, and means "that which cannot be divided." However, with machines called particle accelerators, scientists can smash atoms to pieces, creating even smaller, subatomic particles.

WHAT IS INSIDE AN ATOM?

An atom is made up of three main particles: protons, neutrons and electrons. Protons and neutrons are found at the center, or nucleus, of the atom, with electrons circling around them. Protons have a positive (+) electric charge; neutrons have no charge; and electrons have a negative (–) charge. Sets of electrons circle at different distances from the nucleus, forming electron layers called shells. An atom is mostly empty space: if an atom were the size of a cathedral, its nucleus would be the size of a child's fist. In a hydrogen atom, a single electron circles the nucleus.

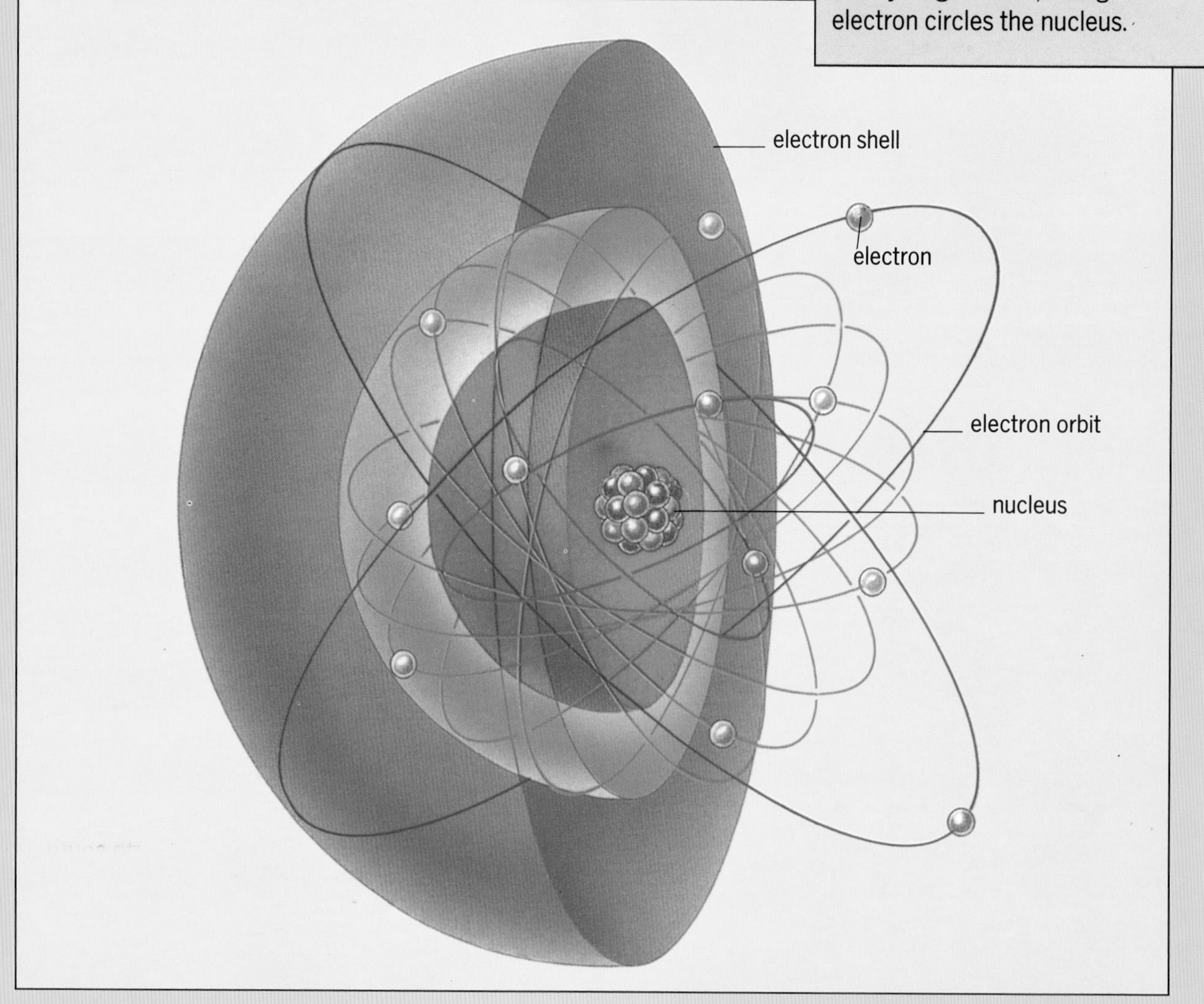

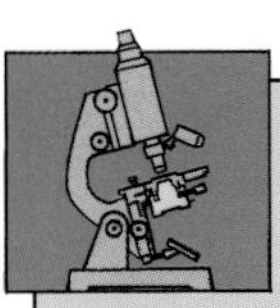

What is a molecule?

Molecules are formed when atoms join together. Most molecules are made up of small numbers of atoms, but many contain more. A molecule of hemoglobin, a substance found in blood, contains 758 carbon atoms, 1,203 hydrogen atoms, plus oxygen, nitrogen, iron and sulphur atoms. A polyethylene molecule has between 12,000 and 60,000 atoms in its molecule.

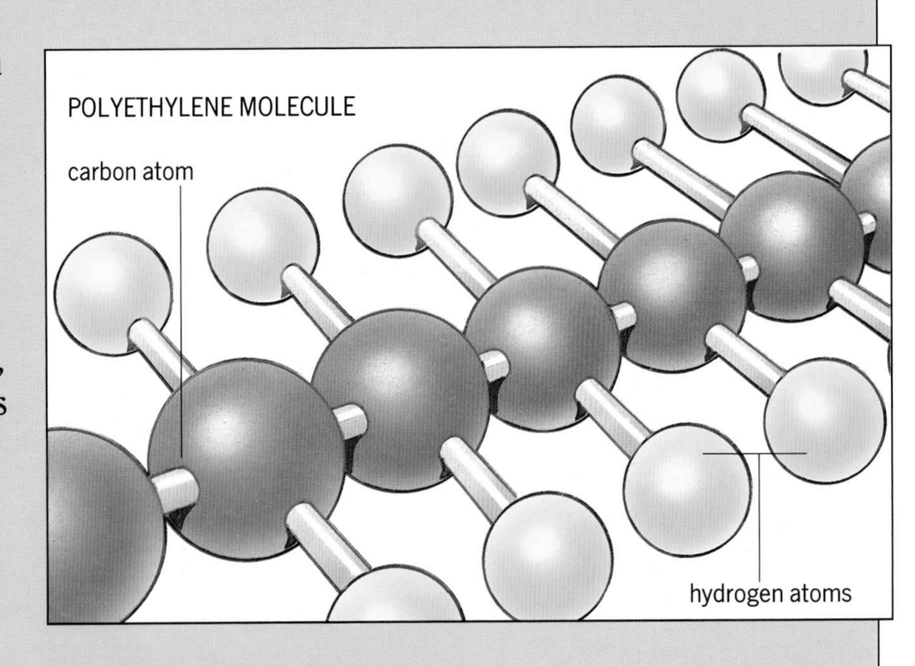

DID YOU KNOW...

- Normally, there are as many electrons circling an atomic nucleus as there are protons in the nucleus? This means that the atom has no overall electric charge.
- Each chemical element has a different number of protons and electrons in its atoms? Hydrogen atoms consist of one proton and one electron. Carbon atoms have six protons and six electrons.
- Atoms with two or eight electrons in their outer shell are particularly stable, or difficult to disrupt? For example, the gas helium has eight outer electrons in its atoms and is "inert"; it does not react easily.
- Atoms can share or exchange electrons in order to form a stable outer shell of eight electrons? This happens when atoms combine to form molecules.

What is inside a proton?

Protons and neutrons are thought to be made up of smaller particles called quarks. Evidence for quarks has been found using particle accelerators, like the one shown under construction below. Other atomic particles seen in accelerator experiments include the neutrino, the tau and the muon. None of these are made up of quarks. Nor do electrons consist of quarks.

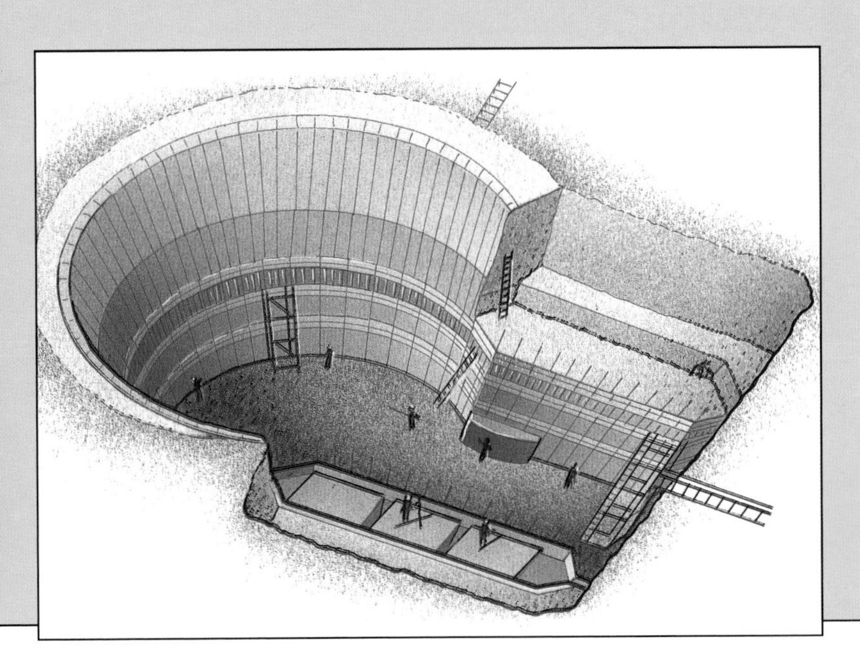

What is a crystal?

Sugar in a sugar bowl is made up of thousands of tiny transparent cubes. These are sugar crystals. A crystal is a small piece of material with flat surfaces and straight sides. Minerals, the chemical compounds that make up the rocks in the Earth's crust, usually occur as crystals. There are many crystal shapes, but each mineral usually forms only one shape, no matter where it is found. The shape of a crystal is determined by the way the particles in the mineral – the atoms and molecules – are packed together.

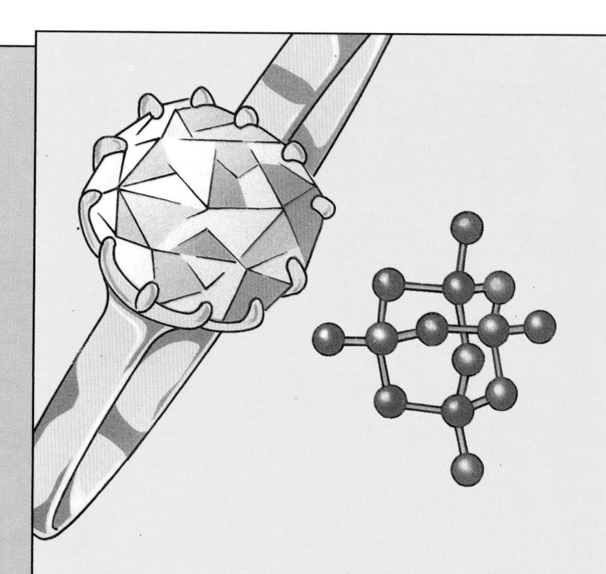

WHAT IS A DIAMOND?

Diamond is a crystalline form of carbon. It is the hardest substance there is. The reason for its strength is that its atoms are strongly linked together and are arranged in a geometric structure that is very difficult to break.

WHAT SHAPE IS A DIAMOND?

Uncut, or natural, diamonds usually have eight flat sides. They can be cut into many different shapes to make gemstones. The size of a diamond is measured in carats. One carat equals about .007 oz (0.2 g). The largest diamond ever found weighed 3,106 carats.

WHAT IS QUARTZ?

Quartz is one of the most common minerals, making up 12 percent of the Earth's crust. It occurs in many different rocks, such as sandstone and granite. It is a form of silica, a compound of silicon and oxygen. Quartz forms pencil-like hexagonal (six-sided) crystals, with a pyramid shape at the end.

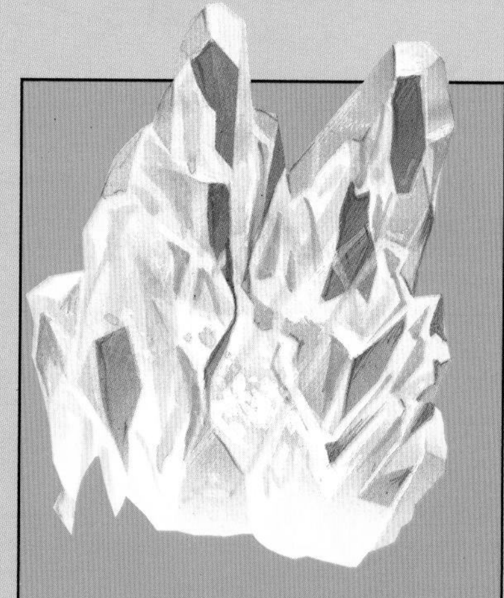

WHAT IS CALCITE?

Calcite is the same chemical compound as chalk. It is the main constituent of limestone rocks. Calcite occurs as pencil-like hexagonal crystals, similar to quartz, called dog-tooth calcite. Crystals over 3 ft (nearly 1 m) across have been found in Missouri and Oklahoma.

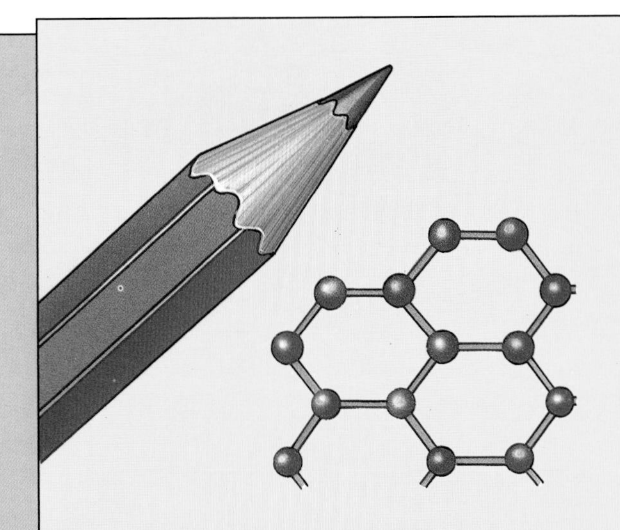

WHAT IS GRAPHITE?

Graphite is a second crystalline form of carbon. It is one of the softest substances known. Its atoms are arranged in flat sheets and there are only weak links between the sheets. Graphite is a weak material and is used as a lubricant.

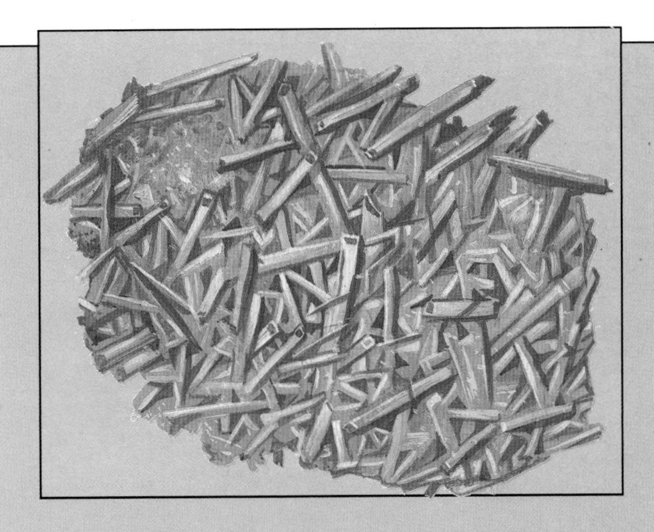

WHY DO CRYSTALS VARY IN COLOR?

The color of a crystal depends upon the chemical composition of the crystal. Many compounds of chromium are brightly colored, like the crystals of crocoite (above). Some crystals are colored because they contain impurities. Quartz, for example, is transparent when it is pure but can have many different colors when impure.

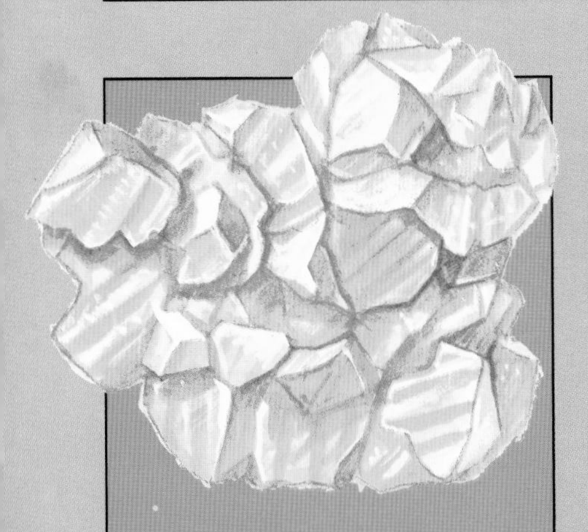

DOES SULFUR FORM CRYSTALS?

Yellow sulfur, found near volcanoes, forms crystals of two different shapes. One form, called monoclinic sulfur, has long flat crystals. The other form, called rhombic, is shaped like two pyramids joined at the base. When rhombic sulfur is heated it becomes monoclinic.

WHAT FORMS CUBE-SHAPED CRYSTALS?

Galena is a compound of lead and sulphur. It forms metallic-looking crystals with a cube shape. Cube-shaped crystals are common. Salt crystals are cubes. A crystal of rock salt contains 40,000 trillion atoms in less than a cubic inch, arranged in tiny cubes.

ARE SNOWFLAKES CRYSTALS?

Snowflakes are made up of small ice crystals built up from water molecules, shaped like small Vs. There are many different ways the V-shaped molecules can link together to form crystals. This is why no two snowflakes are the same. But their shape is always based on a hexagon, with six sides. Large snowflakes can be up to 2.4 inches (6 centimeters) across.

How do solids, liquids and gases differ?

Ice and steam are chemically identical to water. Solids, liquids and gases are different only because their atoms and molecules are held together with different strengths. A solid is dense and strong because its molecules are close together and the force between them is strong. However, they can vibrate slightly. In a liquid, the force between molecules is weaker. The molecules can move around, but they have to force their way between other molecules. In a gas, the molecules can move freely. They spread out to fill their container.

CAN ROCKS BECOME LIQUID?
Heating a solid makes its molecules vibrate more rapidly. Eventually, when the temperature is high enough, the molecules break away from their fixed positions. The solid melts and turns into a liquid. Even solid rocks can turn to liquid if heated enough. When a volcano erupts, the red-hot lava that flows down the slopes is molten (liquid) rock from inside the Earth's crust.

WHAT IS MIST?

The mist over a waterfall, such as Niagara Falls shown left, is not a true gas. It is made up of small droplets of liquid water. The gaseous form of water – water vapor – is invisible. Water vapor forms when water boils or evaporates. In both processes, the water molecules gain enough energy to escape from the liquid. The reverse process, called condensation, occurs when water vapor changes into liquid water. Mist forms high in the atmosphere when water vapor meets a layer of cold air and condenses. Mist droplets may combine to form rain.

WHAT IS THE LIGHTEST SOLID?
The lightest solid is a material called Seagel, made in a laboratory from seaweed. Seagel is lighter than air and would float away if it were not held down.

WHY ARE LIQUIDS WET?
When we touch a liquid, the force between its molecules and those of our skin is stronger than the force between the liquid molecules themselves. So the liquid tends to stick to our fingers.

WHY ARE GASES SQUASHY?
A gas is compressible because there are spaces between its molecules. In air, the distance between molecules is 20 times the diameter of each molecule.

WHY ARE STATUES OFTEN MADE OF METAL?
Metals can be easily worked and cast into statues like this bronze mask. This is because the bonds between atoms in a metal are weaker than those in other materials. The metal atoms are held together by electrons that move between the atoms, acting as a kind of glue. However, the electron "glue" is weak and so the metal can be easily bent and hammered into sheets.

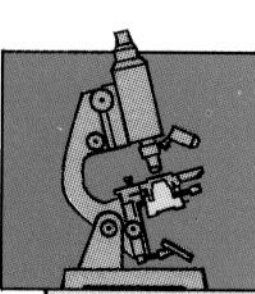

What are elements?

An element is a substance that cannot be broken down into simpler substances. It is made up of a single kind of atom. A compound is a substance that can be separated into different elements. Compounds are made up of two or more elements whose atoms have combined to form molecules. There are 92 naturally occurring elements and 16 artificial elements made in laboratories. Over 10 million chemical compounds have been discovered; 400,000 new compounds are discovered each year.

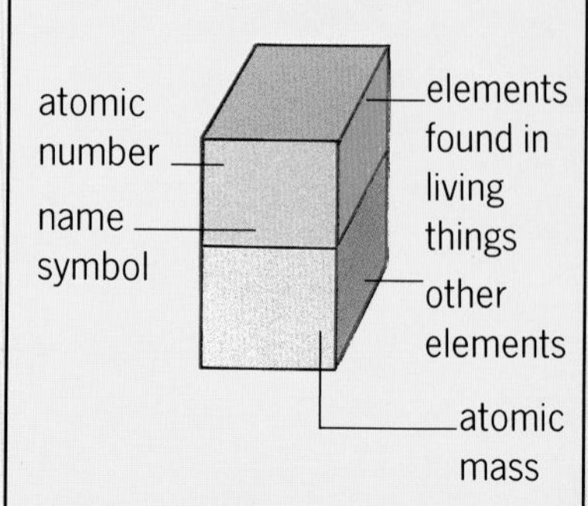

WHAT IS THE PERIODIC TABLE?

It is an arrangement of the chemical elements in which the elements appear in order of their atomic number (the number of protons in their nucleus). The diagram above is a key to the table.

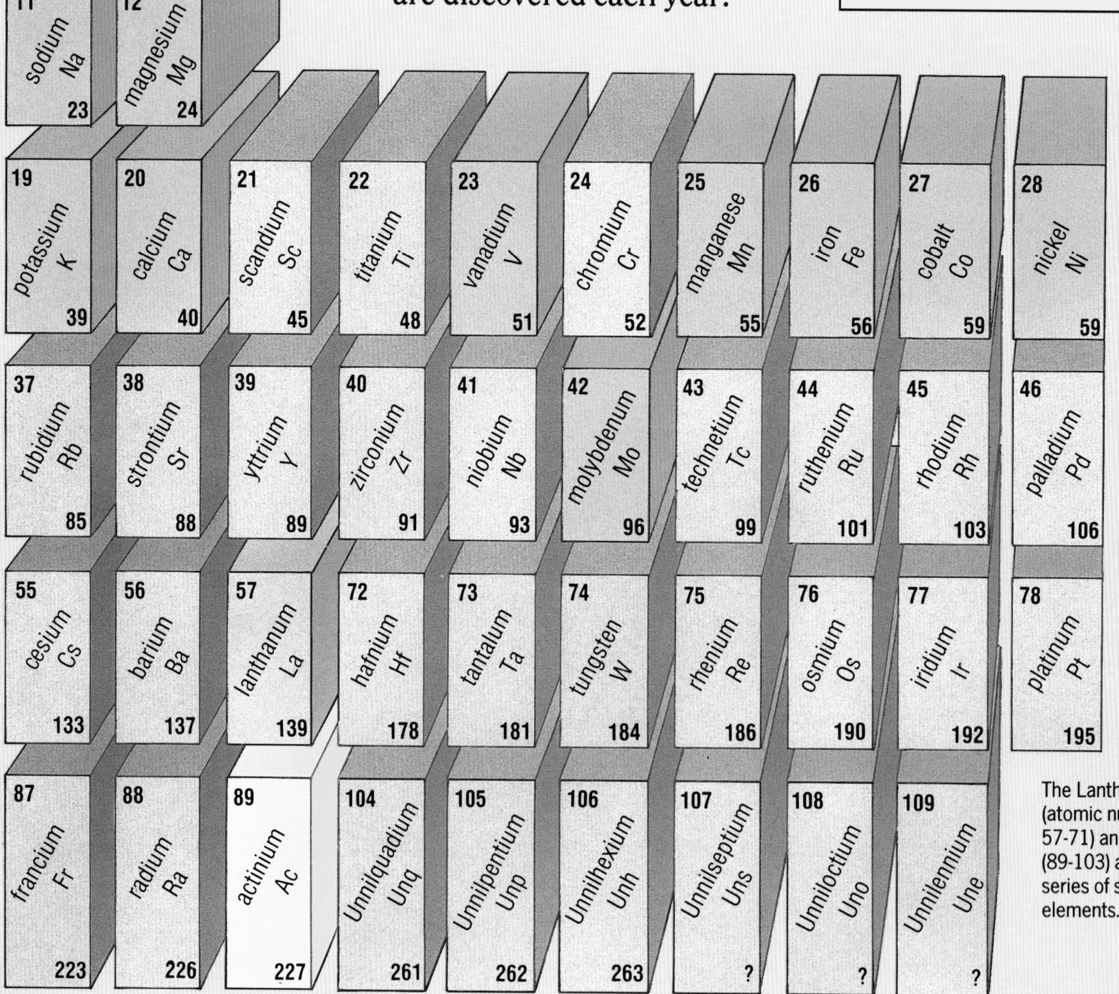

The Lanthanides (atomic numbers 57-71) and Actinides (89-103) are two series of similar elements.

HOW IS THE PERIODIC TABLE ARRANGED?

Elements with similar properties fall in the same vertical column, or group. For example, helium, neon, argon, krypton, xenon and radon, found in the right-hand group of the table, are all very unreactive gases. Elements in each horizontal row, or period, show a gradual change in properties. Moving to the left across the table, the elements gradually become more reactive – magnesium is less reactive than sodium.

The list, right, details the Lanthanides and Actinides (see green and yellow boxes in Table and note on page 30).

NAME	ATOMIC NO.	SYMBOL	MASS
cerium	58	Ce	140
praseodymium	59	Pr	141
neodymium	60	Nd	144
promethium	61	Pm	145
samarium	62	Sm	150
europium	63	Eu	152
gadolinium	64	Gd	157
terbium	65	Tb	159
dysprosium	66	Dy	163
holmium	67	Ho	165
erbium	68	Er	167
thulium	69	Tm	169
ytterbium	70	Yb	173
lutetium	71	Lu	175
thorium	90	Th	232
protactinium	91	Pa	231
uranium	92	U	238
neptunium	93	Np*	237
plutonium	94	Pu*	242
americium	95	Am*	243
curium	96	Cm*	245
berkelium	97	Bk*	249
californium	98	Cf*	249
einsteinium	99	Es*	251
fermium	100	Fm*	253
mendelevium	101	Md*	256
nobelium	102	No*	253
lawrencium	103	Lw*	257

* = artificial element

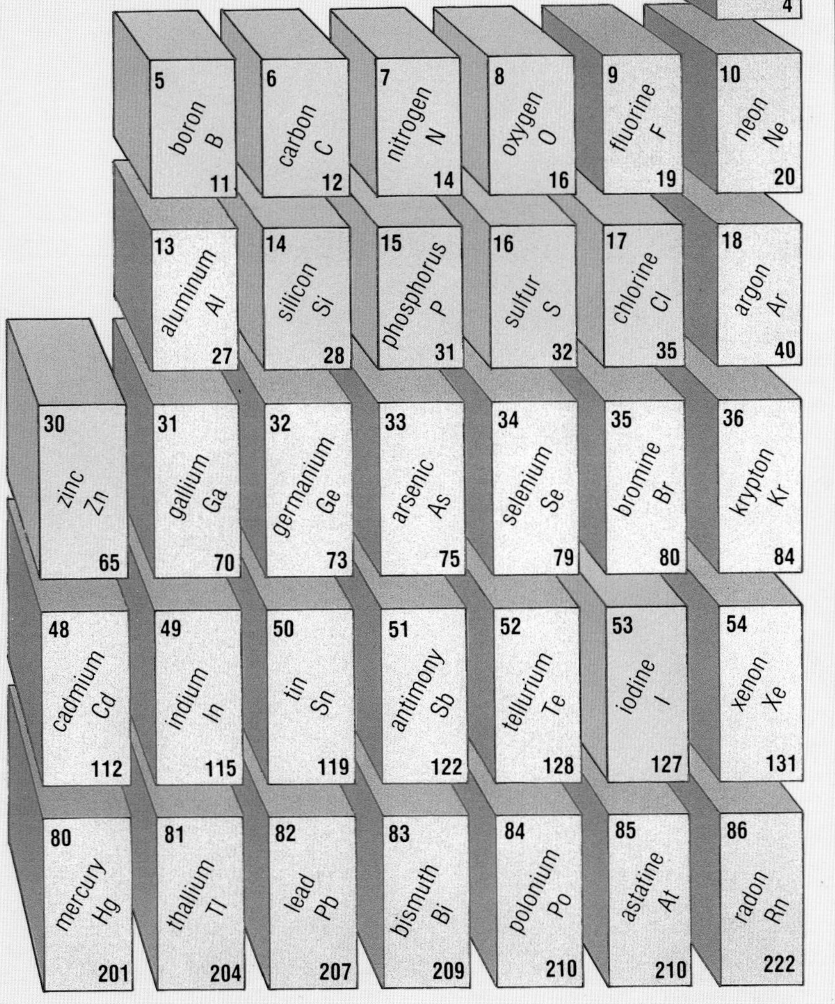

WHICH IS THE MOST COMMON ELEMENT?

The most common element in the universe is hydrogen. Most stars are made up of about 90% hydrogen and 10% helium. Five elements make up over 90 percent of the Earth's crust: oxygen (47%), silicon (28%), aluminum (8%), iron (5%) and calcium (4%). Much of the silicon and oxygen are combined as sand. Three elements make up over 90 percent of all living things: oxygen (62%), carbon (20%) and hydrogen (10%). Four elements make up 99% of seawater: oxygen (85%), hydrogen (11%), chlorine (2%), and sodium (1%).

WHICH IS THE RAREST ELEMENT?

The rarest naturally occurring element is astatine. There is only .012 oz (one-third of a gram) in the Earth's crust.

WHAT ELEMENTS ARE LIQUID?

Of the 92 naturally occurring elements, two are liquids at ordinary temperatures: bromine and mercury.

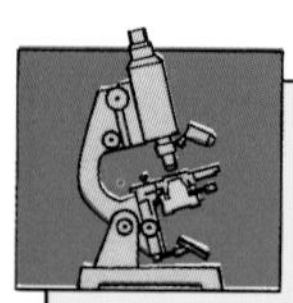

What is a living cell?

All living organisms are made up of cells. The smallest living things have just one cell, but large organisms, like humans, are made up of millions of cells. A typical cell consists of an outer skin, or membrane, containing a liquid called the cytoplasm. The cell's control center, the nucleus, sits in the cytoplasm. There may be small droplets of fluid, called vacuoles, in the cytoplasm. There are also several small bodies, of different types, called organelles. Each type of organelle has a specific role.

DID YOU KNOW...

- Cells measure between .0004 and .002 in (0.01 and 0.05 mm) across?
- A cell may contain hundreds of mitochondria? All the mitochondria in the human body would, if laid end to end, stretch around the world 2,000 times.
- The largest cell of all is the ostrich's egg. It is the size of a small melon, about 6 in (150 mm) across.

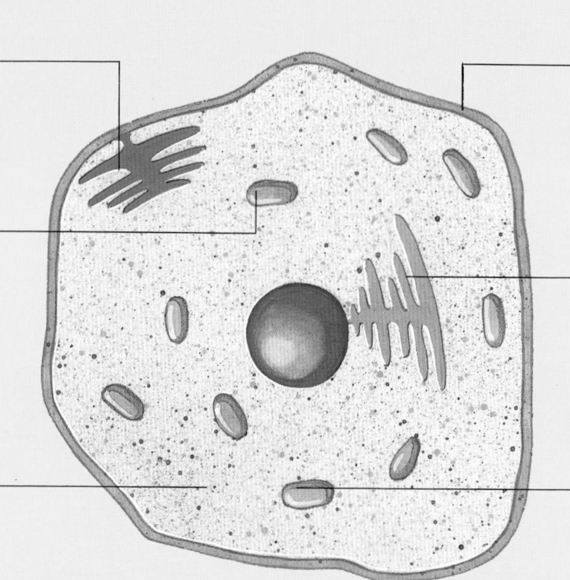

Golgi body. Waste materials are collected in these small sacs, and carried out of the cell.

Mitochondria. The cell's energy packs. In these organelles simple fuel chemicals are broken down to provide energy.

Cytoplasm. The cytoplasm generally has a jelly-like outer layer and a liquid center.

Cell membrane. The outer skin of the cell. Chemicals can flow in and out of the cell through the membrane.

Endoplasmic reticulum. A twisting passageway between the cell membrane and the nucleus. Fluids pass through it.

Lysosomes. Round sacs that take in foreign bodies, such as bacteria, to be destroyed.

ANIMAL CELL

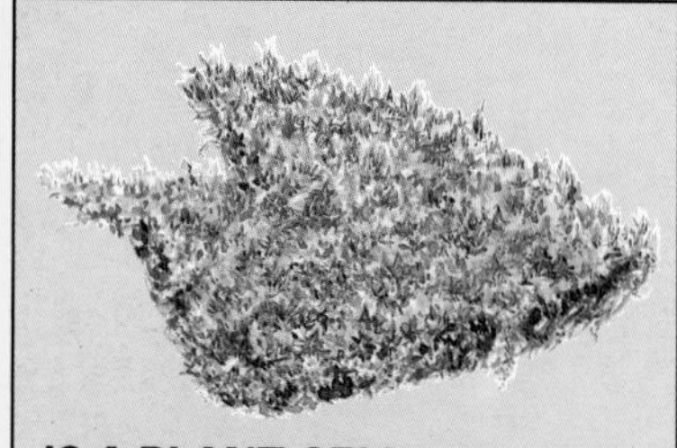

IS A PLANT CELL DIFFERENT?

A typical plant cell contains chloroplasts, small bodies that absorb sunlight during photosynthesis and make food. Also, plant cells have a cell wall made of cellulose.

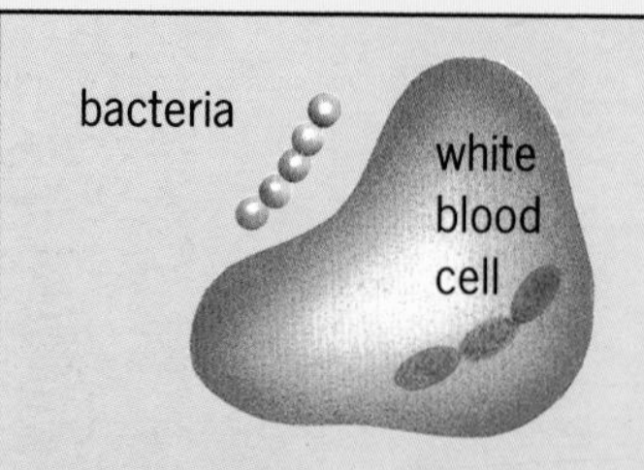

WHAT ARE BACTERIA?

Bacteria are single-celled organisms between .00002 and .0002 in (0.0005 mm and 0.005 mm) in size. About 1,000 of them could fit on the period at the end of this sentence.

WHAT ARE SPORES?

Fungi, such as mushrooms and molds, use special cells, called spores, to reproduce. A single giant puffball may contain 7 billion spores, which are released to the wind for dispersal.

What is an amoeba?

Not all single cells are simple. An amoeba is a single-celled creature that is more complex than any single cell found in the human body, having all the organelles and chemical apparatus needed to move about and lead an active life. The flexible amoeba cell extends in one direction, as though the creature was extending a leg forward. The rest of the cell flows after the "false foot," or pseudopodium.

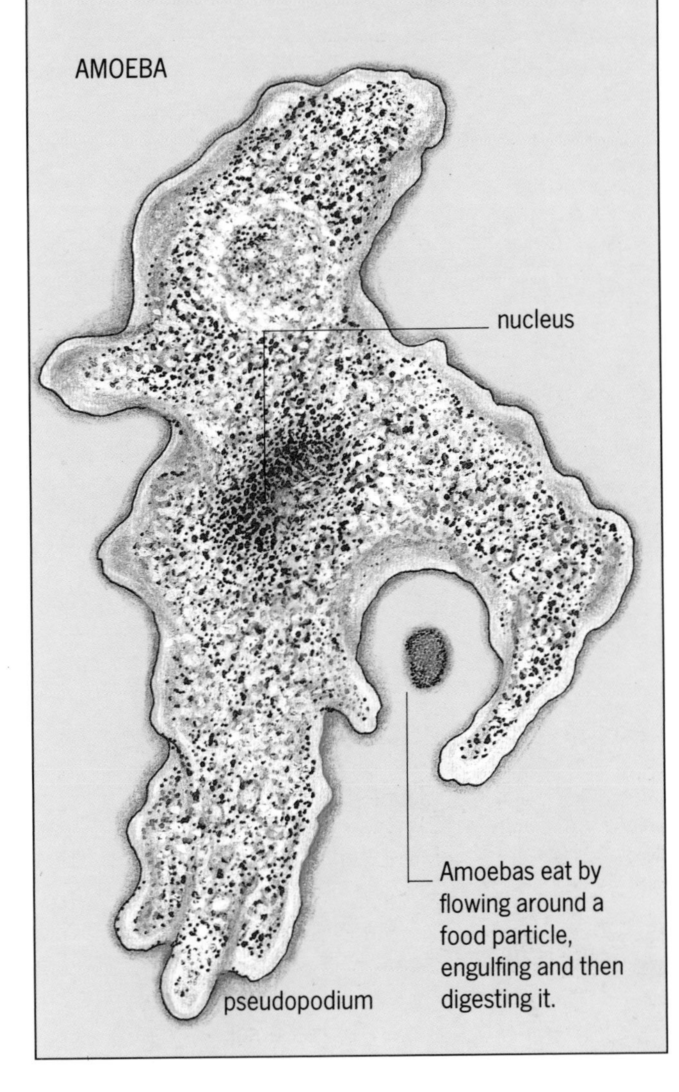

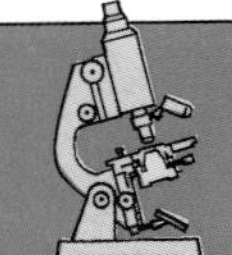

How are bodies built?

In a simple animal, such as a jellyfish, there are just two types of cell. In more complex animals there are many more types. Cells of the same type are arranged in groups called tissues. Muscle tissue, for example, is made up of long muscle cells. Tissues of a particular type are grouped together to make organs; the heart, for example, is made up of layers of muscle tissue. Organs link together to make body systems; the heart is part of the blood circulation system.

What is a lever?

A lever is a simple machine for lifting weights or applying a force. It consists of a strong, stiff bar that turns around a pivot, or fulcrum. A lever can magnify a force. For example, a small force (effort) applied to the end of a bottle opener produces a force (load) on the bottle cap strong enough to lift the cap. Seesaws and nutcrackers are also levers.

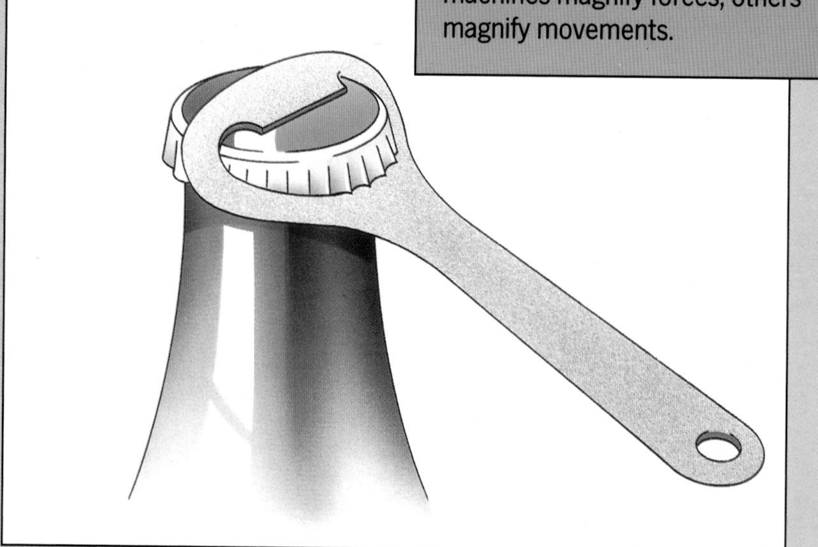

WHAT IS A MACHINE?
Machines are devices that apply forces to carry out useful tasks such as lifting heavy objects. Some machines magnify forces, others magnify movements.

How do levers magnify a force?

On a seesaw, a person weighing 110 lbs (50 kg) is balanced by a person weighing only 55 lbs (25 kg) who sits twice the distance from the fulcrum. Using a ruler as a lever, and an eraser as a fulcrum, you need little effort to lift a book, but the book must be closer to the fulcrum than your point of effort.

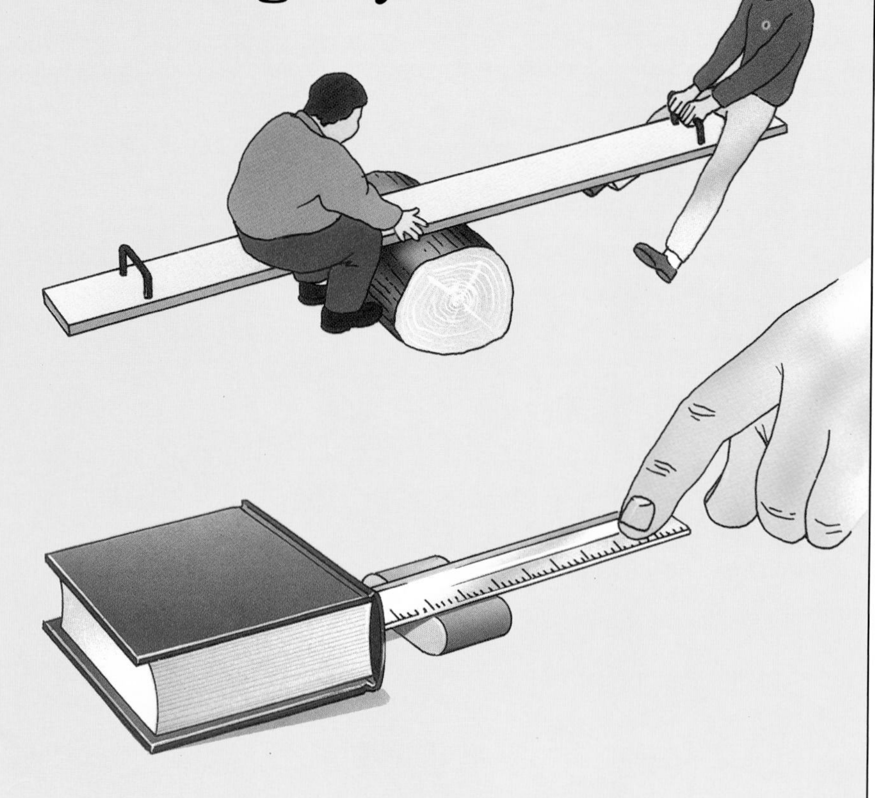

DID YOU KNOW...
On a piano keyboard, levers make the hammers that hit the strings travel ten times the distance moved by the keys.

Are there many types of lever?

Yes. Levers differ according to where the fulcrum is and where the effort and load (force produced) act. The crowbar is a lever in which the effort and load act on opposite sides of the fulcrum. With a wheelbarrow, the fulcrum is at one end, the load is in the middle and the effort is applied at the handles. Nutcrackers are a pair of levers sharing a fulcrum at one end. Sugar tongs are levers with the effort between the load and the fulcrum.

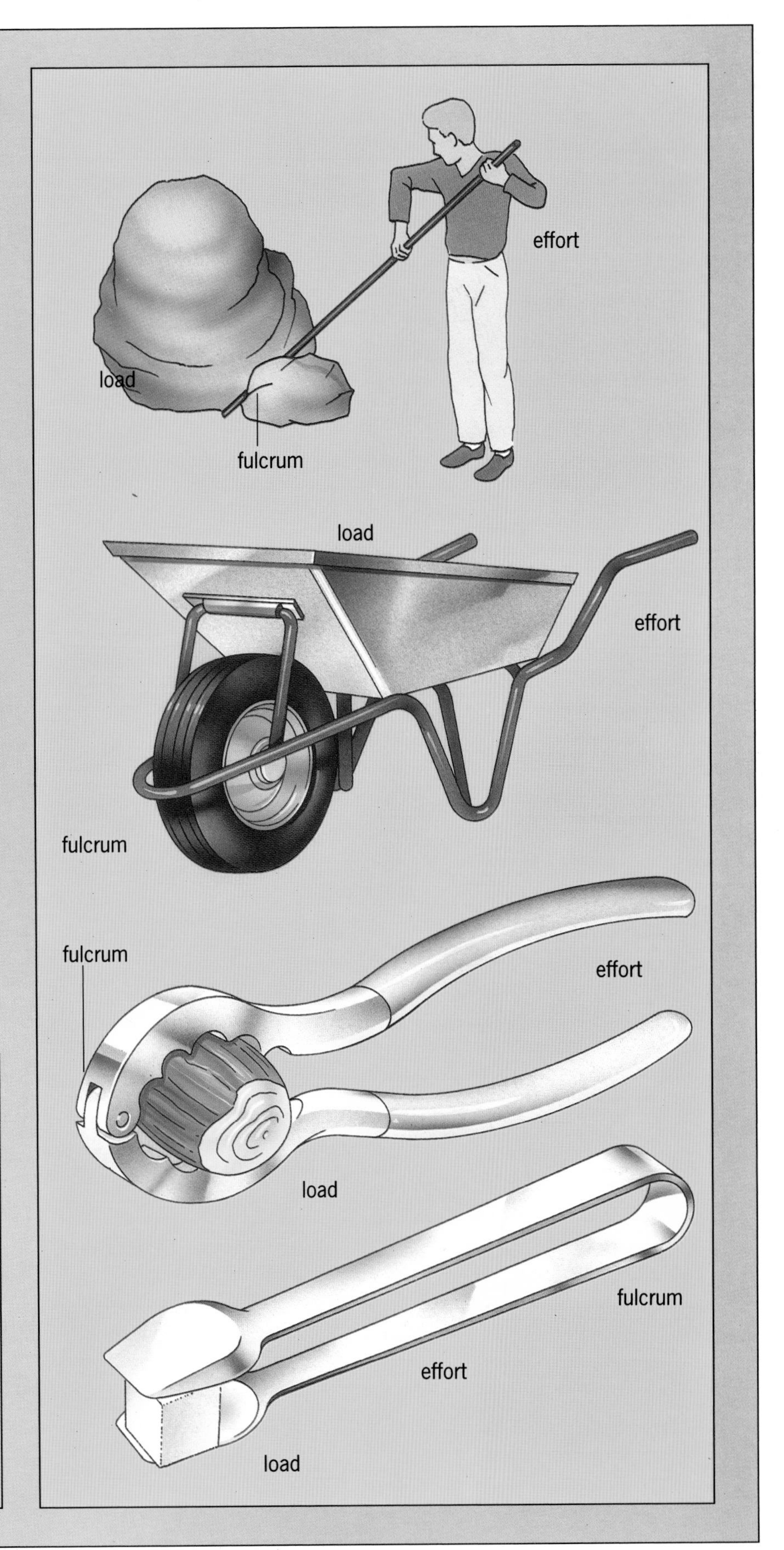

DID YOU KNOW...

The ancient Greek scientist Archimedes said, "Give me a long enough lever and a place to stand, and I could move the Earth"? He realized that a long crowbar, with the load close to the fulcrum, could lift very heavy weights. In 212 B.C., when a Roman fleet attacked Syracuse, in Sicily, Archimedes designed huge machines that lifted the Roman ships from the water using levers.

What is a pulley?

A simple pulley is a rope stretched over a single wheel. Simple pulleys change the direction in which a force acts, but do not magnify the force. A pulley with more than one wheel – a compound pulley – is able to magnify a force. Like a lever, a compound pulley allows a small effort to overcome a large load because the effort moves through a greater distance than the load. Using a pulley with four wheels, one person can lift a weight that normally needs four people to lift. But the effort will move four times as far as the load.

HOW IS A RAMP LIKE A MACHINE?

A ramp (sometimes called an inclined plane) can be considered a simple machine. Walking up a slope, or carrying a load along a slope, is much easier than climbing or lifting vertically. Of course, the distance moved is greater when walking up a slope. If the length of the slope is, say, 33 ft (10 m) and the vertical height to the top is 16.5 ft (5 m), the effort needed to move a load up the slope is half that needed to raise it up the vertical side. Screws are also simple machines; their thread is like an inclined plane wrapped around an upright column.

PULLEYS AT WORK

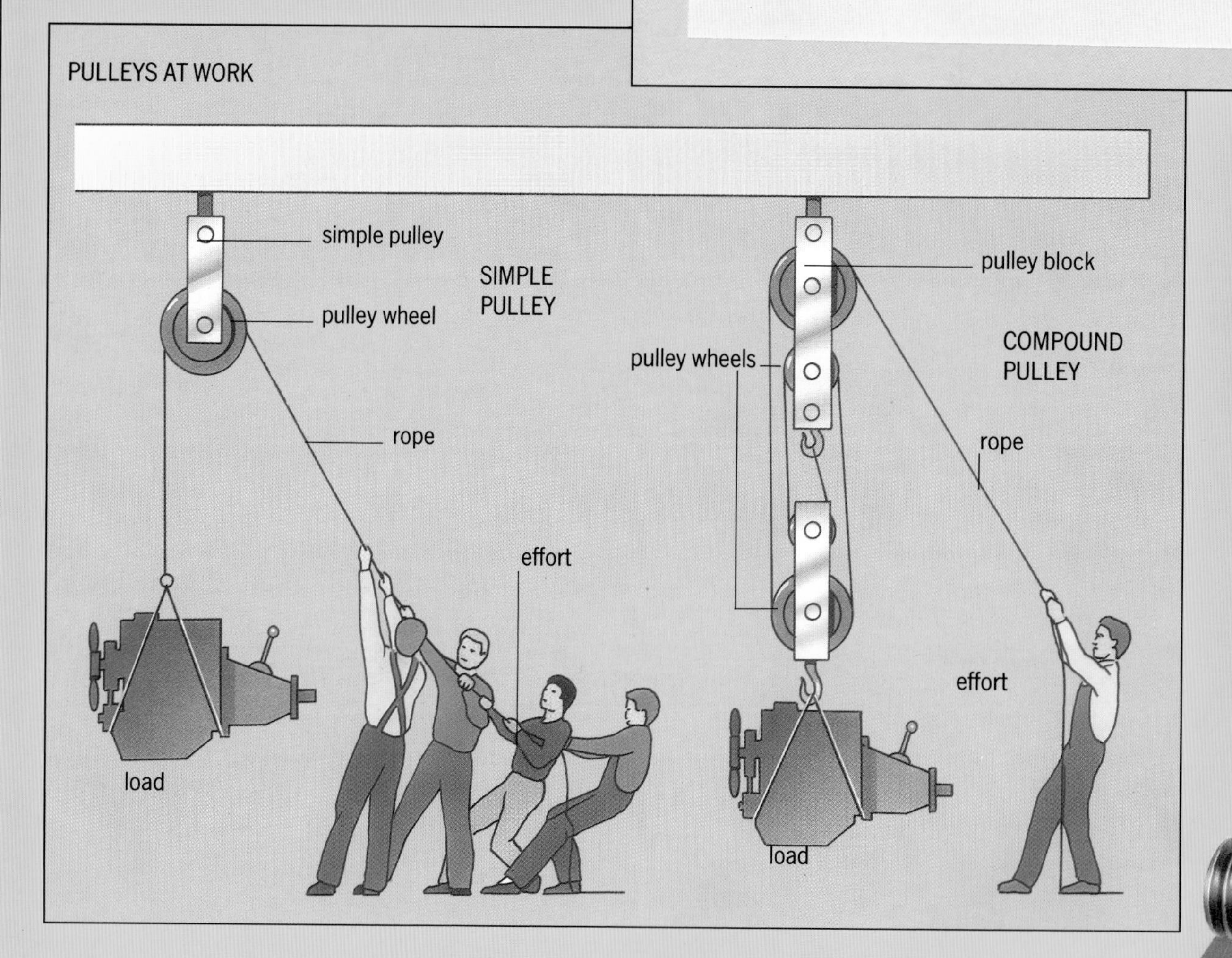

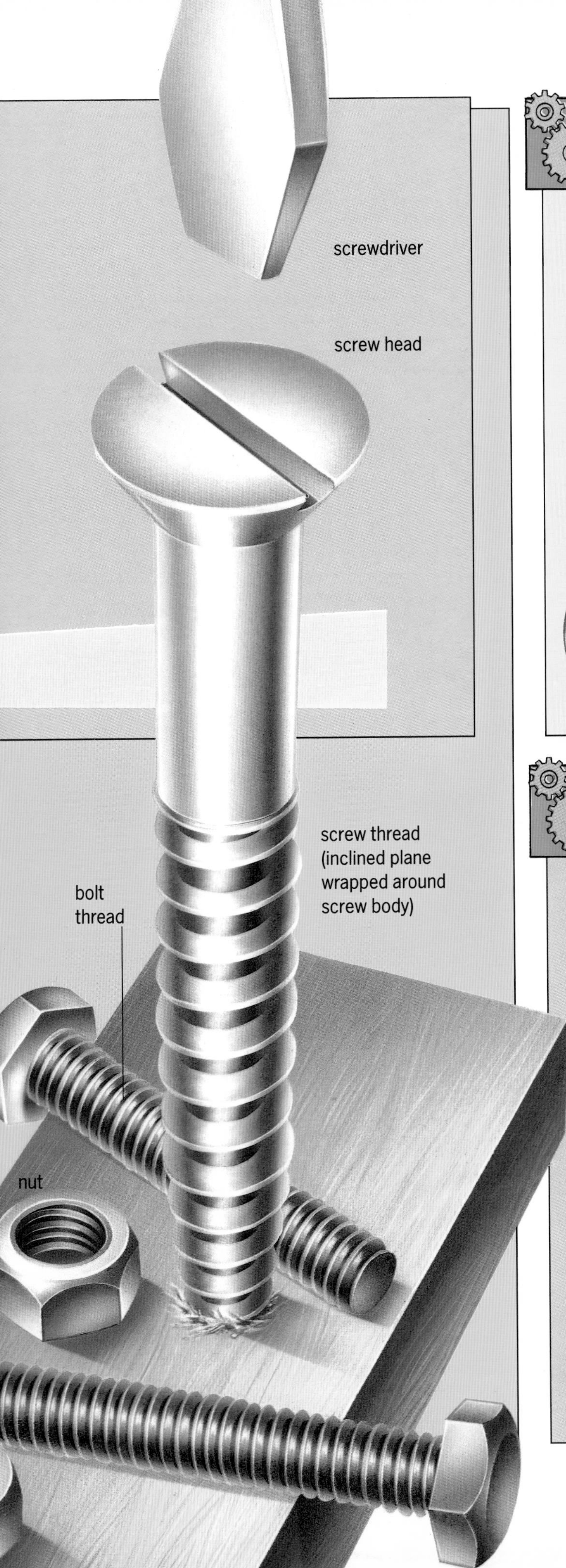

How does a wrench work?

When a wrench is turned by a small force applied to its end, a larger force is generated at its jaws – enough to tighten a bolt. As in any machine, the small force moves a greater distance than the large force when the wrench turns.

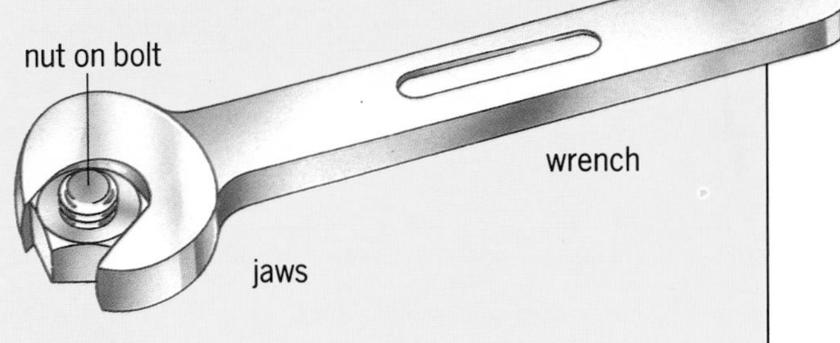

Is a brace and bit a machine?

The curved handle of a brace and bit is like a wheel and axle, or a wrench. A small force on the handle turns the bit with a greater force. The turning effort is also magnified by the screw-shaped grooves on the bit. The bit moves forward with a greater force than the effort used to turn it.

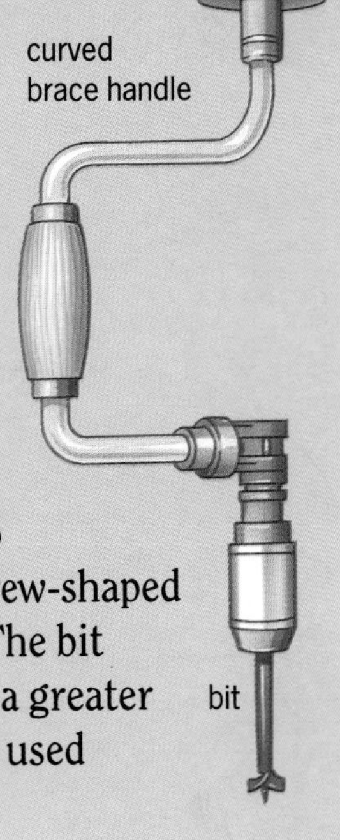

How does an excavator work?

A mechanical digger consists of many simple machines – levers, screws, pulleys and gears – linked together by belts, chains, transmission shafts and hydraulic pipes. The digging arm of an excavator is a linked series of levers, for example. Power is transmitted to the digging bucket by liquid, or hydraulic, pressure.

DID YOU KNOW...

- Probably the most complex mechanical device is a large clock in the Cathedral of St. Pierre, Beauvais, France? It has 90,000 parts, and is 40 ft (12 m) high, 20 ft (6 m) wide and 9 ft (2.7 m) long.
- The tallest mobile crane in the world is 667 ft (202 m) high, as tall as a 60-story building? It can lift 33 tons to a height of 528 ft (160 m). It is carried on 10 huge trucks.

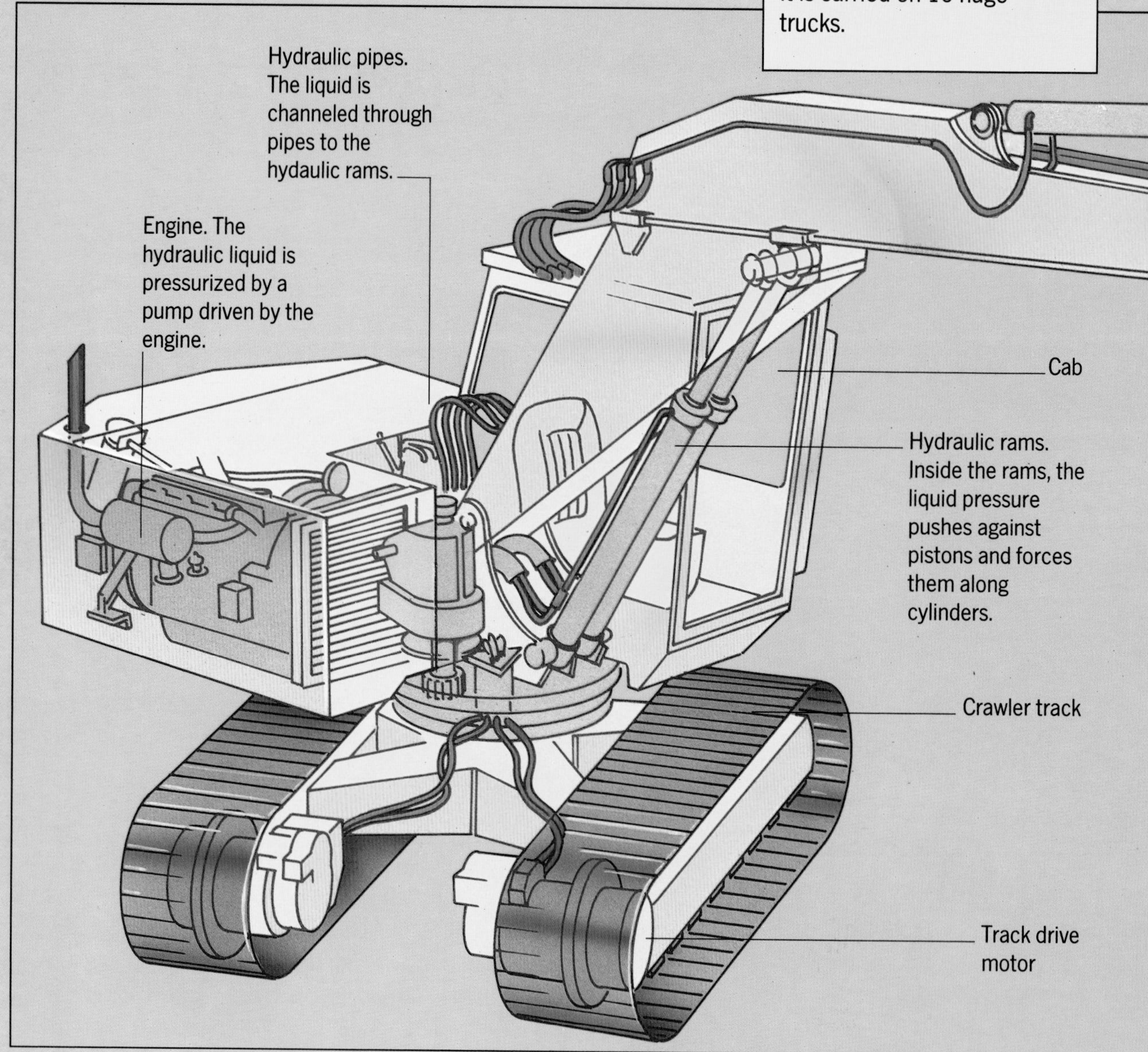

HOW DOES A CAROUSEL WORK?

A fairground carousel or merry-go-round is driven by gear wheels. Gears are sets of interlocking toothed wheels. They can increase or decrease the strength and direction of a force, and change the speed and direction of a rotation. Cars, buses and trucks are fitted with sets of gears.

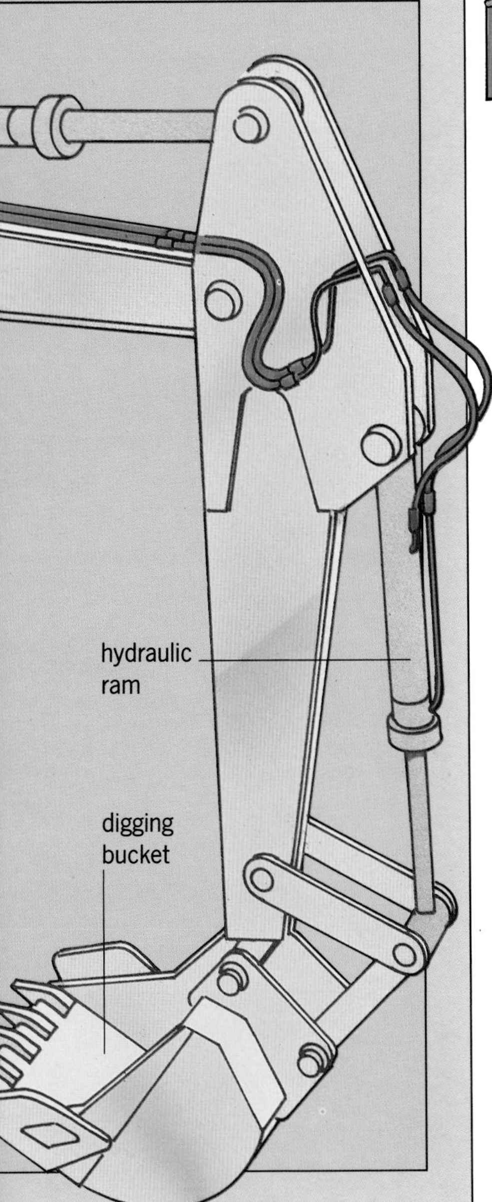

What powers a plane?

All machines need a source of energy. In an airplane, the energy comes from the engine fuel. The engine converts the chemical energy of the fuel into the energy of movement. In a propeller airplane, fuel is burned in the engine cylinders and the gases produced push against pistons. The movement of the pistons turns the propeller via levers and gears. Jet planes work in a similar way.

What is an electromagnet?

A coil of wire carrying an electric current has a magnetic field like a bar magnet. The stronger the electric current, the stronger the magnetic force produced. These magnets are electromagnets. Their magnetism can be switched off and on by switching the current off and on.

In junkyards, powerful electromagnets are used to lift scrap cars. The electromagnets can lift a car weighing more than a ton.

WHAT MAKES A BELL RING?
When the button is pressed, current flows through wire coils turning them into electromagnets. The steel or iron armature (bar) is attracted and moves toward the coils. This makes the clapper strike the bell. The movement breaks the electrical circuit as the contact points separate. The current stops and the armature returns to its original position.

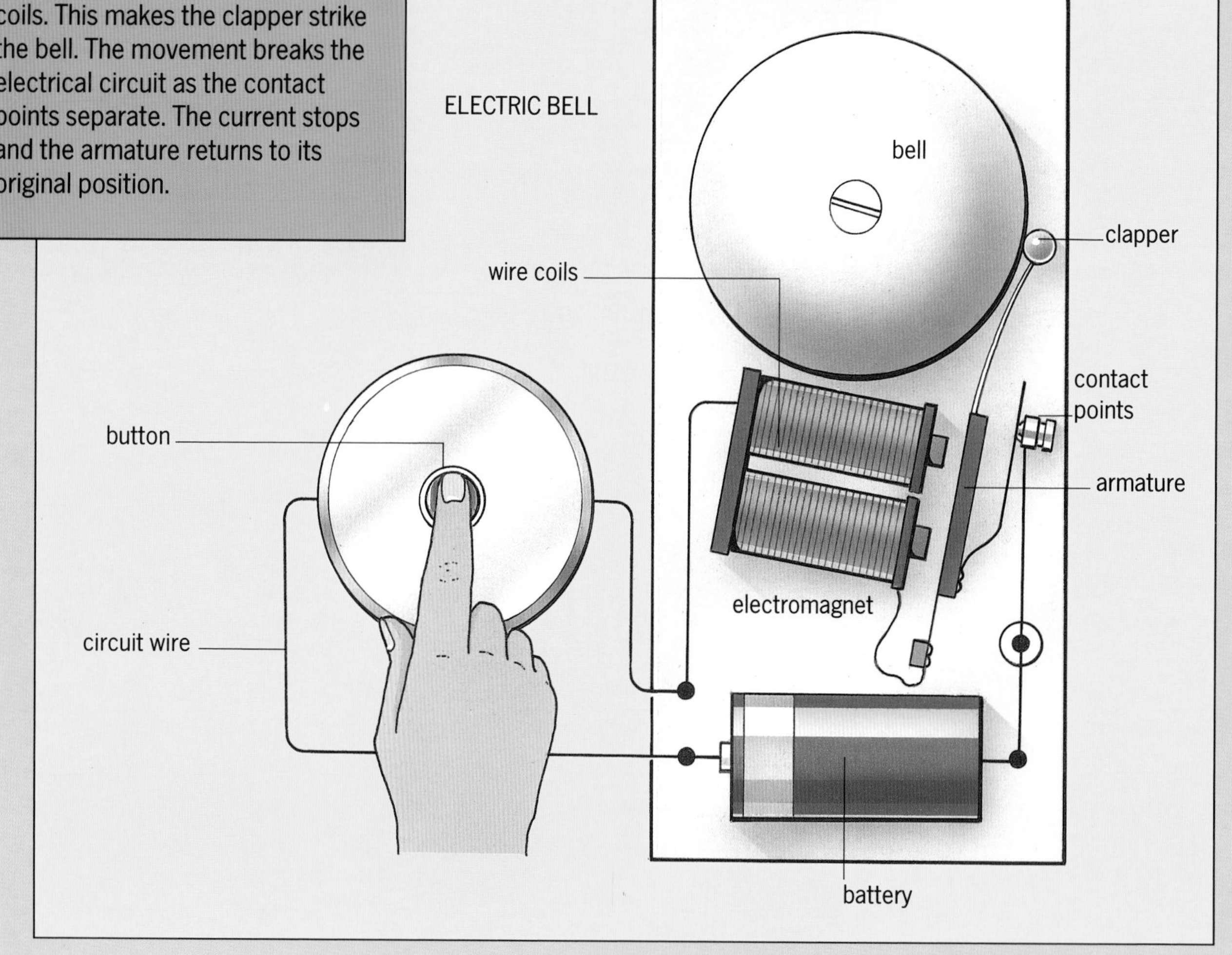

ELECTRIC BELL

How do batteries work?

Batteries produce electricity by means of chemical reactions. In a dry-cell flashlight battery, reactions take place in a chemical paste between two terminals or electrodes when these are connected by a wire or circuit.

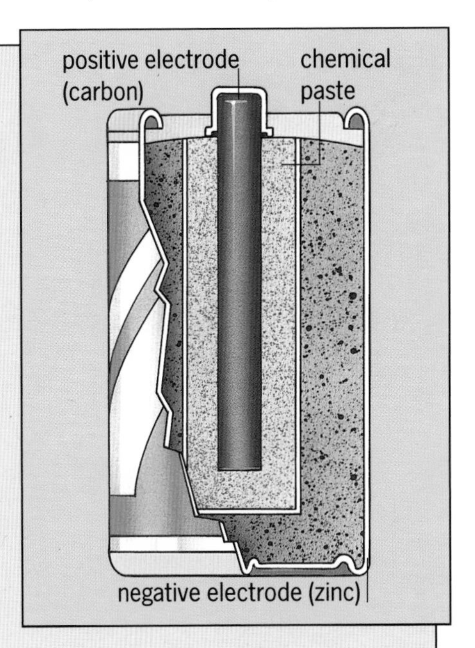

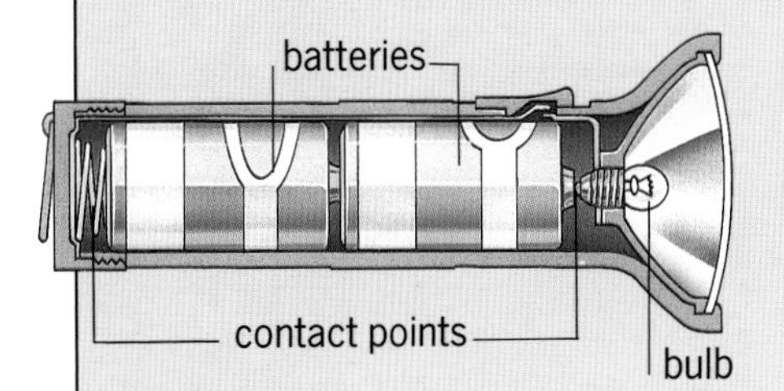

In this flashlight, one terminal of the bulb is in contact with the top battery. Pressing the switch connects the bottom battery and the other terminal of the bulb.

How does an electric motor work?

Current from a battery passes through the field coils, turning them into electromagnets. The armature coils also become magnets as current flows through them. They are attracted by the field coils and forced to rotate.

DID YOU KNOW...

- About 3 trillion electrons flow through the filament of a light bulb every second?
- An electric car broke the 60 mph (100 km/h) barrier in 1899, at Achers near Paris?
- The most powerful electric current in the world is produced at the Los Alamos Scientific Laboratory, New Mexico? For a fraction of a second, as much current is generated as everywhere else on Earth put together.

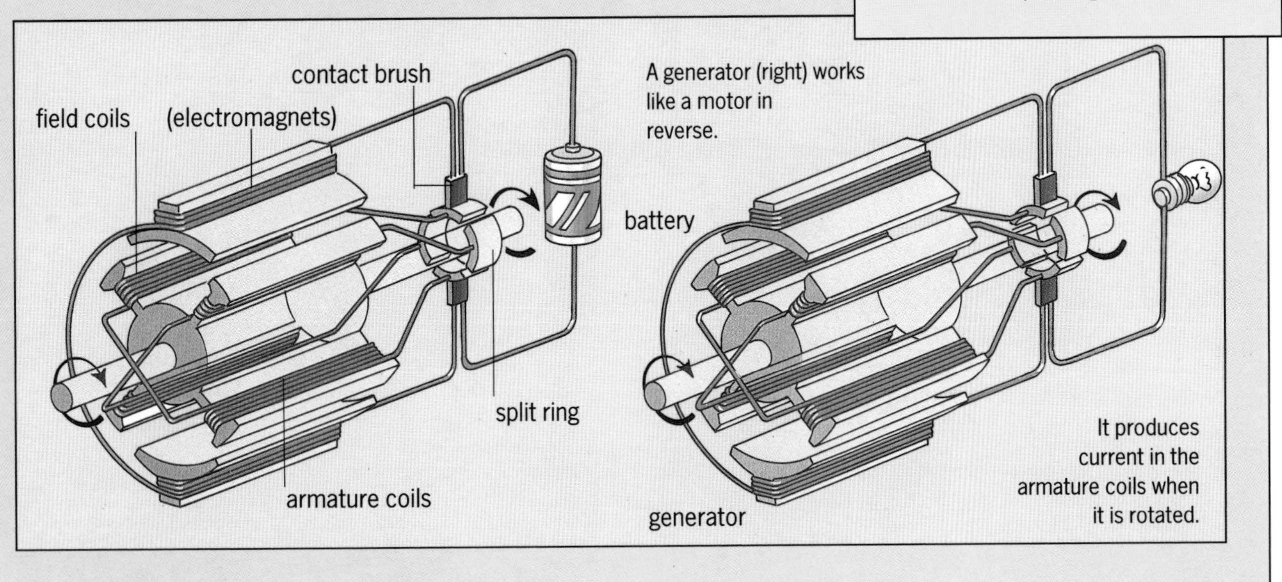

A generator (right) works like a motor in reverse.

It produces current in the armature coils when it is rotated.

What is temperature?

We measure temperature in degrees (°) using a thermometer (right). The temperature of an object is a measure of its hotness or coldness, which is related to how rapidly its atoms and molecules are vibrating. The hotter an object, the faster its particles vibrate.

There is one temperature at which atoms and molecules stop moving: −459.67°F (−273.15°C). It is impossible to cool things below this temperature, called absolute zero.

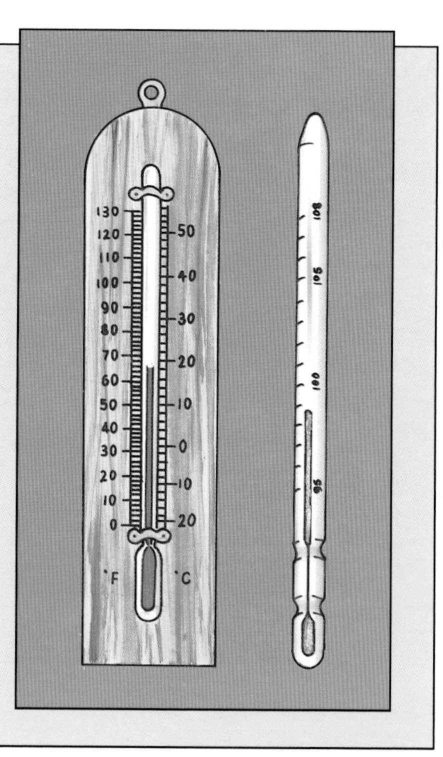

DID YOU KNOW...
Crocus flowers are natural thermometers? They open when the temperature rises and close when the temperature drops, reacting to temperature differences of only 0.9°F.

Is heat the same as temperature?

No. The heat in an object is the total amount of movement energy of its molecules. A small object that is very hot does not hold as much heat as a large object that is not so hot. A large iceberg contains enough heat to boil water, if only the heat could be collected and concentrated.

WHY DOES AN ICEBERG FLOAT?
An iceberg floats because water expands as it cools below 39°F (4°C) and freezes. This means that an iceberg is less dense than the water around it. This behavior is unusual; most substances get smaller as they get colder. In winter, when a pond starts to freeze, the coldest water rises to the top. Fish can live in the warmer water near the bottom even when the surface is frozen.

Why do things expand?

When a solid object is heated, its molecules move more rapidly and move farther away from their normal position. This makes the object get larger. A 3,300-ft- (1,000-m-) long bridge can extend 20 in (50 cm) on a hot summer's day. Gases and liquids also expand when heated. A hot air balloon rises because the heated air inside has expanded and become lighter than the colder air around it.

DID YOU KNOW...

The highest temperature in the universe – 720 million °F (400 million °C) – was produced by Japanese scientists for a fraction of a second? This is 200 times hotter than the surface of the Sun (3.6 million °F, or 2 million °C), 60 times hotter than the center of the Sun (27 million °F, or 15 million °C), and hotter than the hottest star (540 million °F, or 300 million °C). The coldest temperature ever reached was produced at the University of Lancaster, England – one-millionth the temperature of outer space (−454 °F, or −270 °C).

Why are arches used in buildings?

Before the arch was invented about 5,000 years ago, the roof of a big building was supported by beams resting on side walls or columns. With a roof over 23 ft (7 m) wide, the beams were likely to break. The arch solved this problem by carrying the weight of the roof on to the side walls or columns. An arch is made by fitting together stone wedges. The keystone, at the top, is the last to be put in place.

What stops a dam from bursting?

A dam resists the weight and flow of the water behind it. An embankment dam is a thick wall of soil or rock. A gravity dam is a concrete wall whose sheer weight holds back the water. Curved or arch dams are used in deep narrow valleys. As water presses against the dam, the arch is squeezed more tightly against the valley walls.

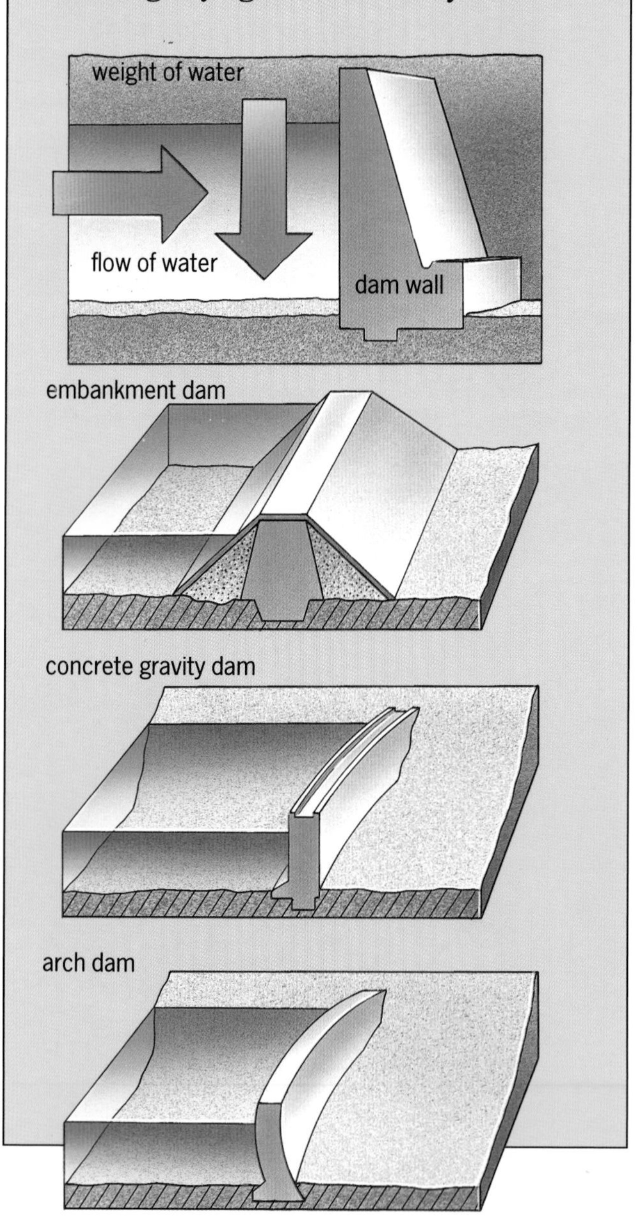

What stops a bridge from collapsing?

A bridge must be strong enough to carry the weight of the traffic crossing it. In a truss or beam bridge, the weight of the bridge and its load is supported by large blocks at each end. The beam may be made of a framework of steel girders.

The weight of a concrete arch bridge pushes outward and downward on the end supports. Suspension bridges use cables to pull outward toward anchorages on either side of the span. The roadway hangs from the main cables.

truss or beam bridge

concrete arch bridge

suspension bridge

WHERE WERE THE FIRST ARCHES BUILT?
Arches were used to span small drains and sewers in Pakistan between 3000 and 2000 B.C. Around 1400 B.C., the Egyptians built arches.

WHERE IS THE LARGEST CONCRETE DAM?
The Grand Coulee is on the Columbia River in Washington State. It is 551 ft (167 m) high and 4,198 ft (1,272 m) long.

WHERE IS THE LONGEST BRIDGE?
The world's longest bridge crosses Lake Pontchartrain, joining Lewisburg and Mandeville, Louisiana. It is almost 23 mi (39 km) long.

What is a flightpath?

Airline pilots fly their craft on set routes called flightpaths. These are marked out by "waypoints," which are points in the sky above ground radio or radar transmitters, called beacons. The aircraft finds its way to a waypoint by tuning in to the beacon radio or radar.

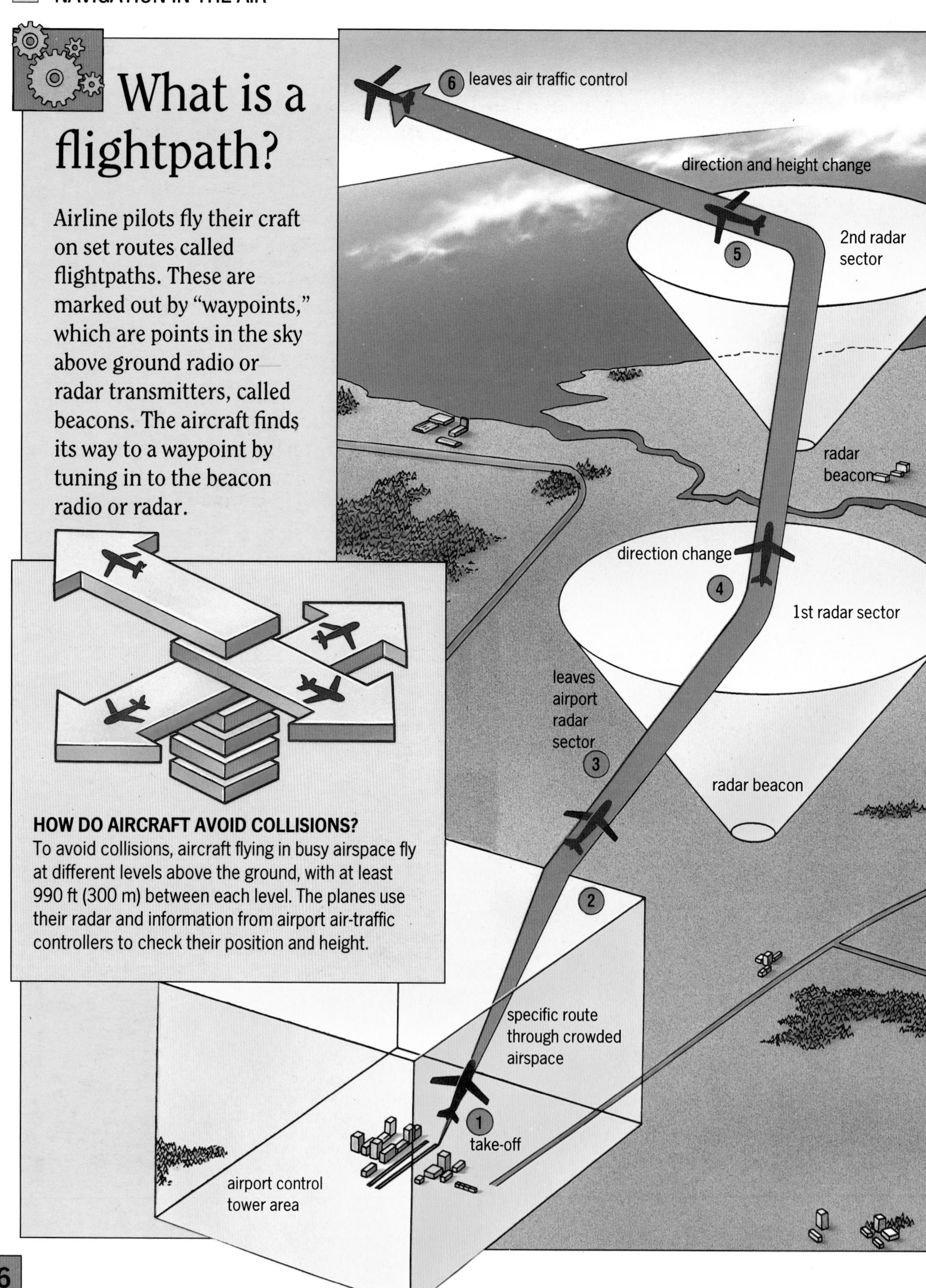

HOW DO AIRCRAFT AVOID COLLISIONS?
To avoid collisions, aircraft flying in busy airspace fly at different levels above the ground, with at least 990 ft (300 m) between each level. The planes use their radar and information from airport air-traffic controllers to check their position and height.

WHAT IS RADAR?

Radar (short for RAdio Detection And Ranging) is a way of finding an aircraft's position using radio echos. Very short pulses of radio waves are sent out by a transmitter. These pulses bounce off any aircraft within range before returning to a receiver. The distance and direction of the aircraft can be worked out from the reflected signal.

What do air-traffic controllers do?

They tell each aircraft what course to follow as it flies through the airport control area. Each aircraft carries radar equipment that continuously transmits to the control center information about its range and height. The airport radar also tracks each aircraft. The signals received at the control center are relayed to a display screen that the controller monitors. Many details are displayed on the controller's screen: aircraft height and speed, rate of climb, destination.

DID YOU KNOW...

- The largest airport in the world is the King Khalid Airport at Riyadh, Saudi Arabia?
- The busiest airport in the world is O'Hare International Airport in Chicago, Illinois? Over 59 million passengers pass through the airport each year. An aircraft takes off or lands every 40 seconds of the day and night.
- Heathrow Airport near London handles more international travelers than any other? In 1988, a record of 12,446 passengers were handled in one hour.
- The world's largest airline is the Russian state airline Aeroflot, which carries more than 132 million passengers in a year?

WHY DO AIRCRAFT CIRCLE ABOVE AIRPORTS?

If too many aircraft arrive at an airport at the same time, they may have to wait their turn to land. Unlike vehicles caught in a road traffic jam, aircraft must keep moving to maintain lift. They are directed by air-traffic control to fly in a tall circular pattern called a stack, until they can land.

The stacking area is marked by a radar beacon. Within the stack, planes fly in a circle or oval, one above the other, with 990 ft (300 m) of height between them. When an aircraft arrives, it enters the top of the stack. When the plane at the bottom of the stack begins its final approach to the runway, each plane above it is told to come down to the next level. At some airports, aircraft in the stack spiral slowly downward to reduce fuel usage and noise.

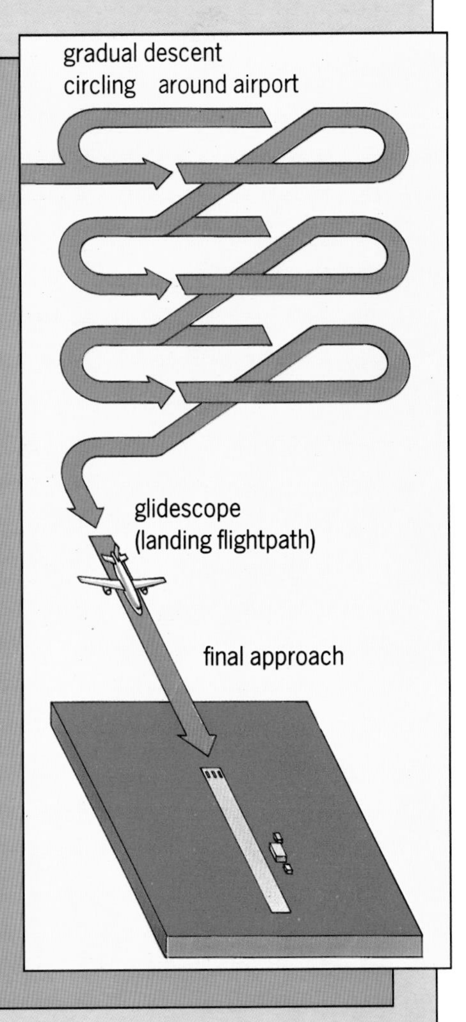

INDEX